FRANK M. POST,

Landscape and Garden Work

PLANS AND ESTIMATES

Choice Home Grown Roses

Every Kind of Tree, Shrub or Vine

21 Henderson Terrace   Burlington, Vt.

from

Dr. H. S. Phelps,

MANUAL OF FRUIT INSECTS

MANUAL

OF

FRUIT INSECTS

BY

THE LATE MARK VERNON SLINGERLAND

AND

CYRUS RICHARD CROSBY

OF THE NEW YORK STATE COLLEGE OF AGRICULTURE
AT CORNELL UNIVERSITY

New York

THE MACMILLAN COMPANY

1914

Norwood Press
J. S. Cushing Co. — Berwick & Smith Co.
Norwood, Mass., U.S.A.

To

WILLIAM SAUNDERS

LEADER IN AGRICULTURAL INQUIRY

AUTHOR OF "INSECTS INJURIOUS TO FRUITS"

WHICH

FOR NEARLY ONE-THIRD OF A CENTURY

HAS BEEN THE STANDARD WORK ON THE SUBJECT

THIS BOOK IS DEDICATED

AS A TOKEN OF APPRECIATION

PREFACE

For nearly twenty years Professor Slingerland, as Assistant Entomologist of the Cornell University Agricultural Experiment Station, devoted the greater part of his time to studying the insect problems encountered by the fruit-growers of New York State. The results of some of these studies were published as bulletins of the Station, but a large amount of material remained unpublished. It was the idea of making this material available to other workers and at the same time bringing together in connected form all the more important known facts concerning the insect enemies of our deciduous fruits, that led Professor Slingerland in the fall of 1908 to begin writing this book. During the few remaining months of his life he worked rapidly, and in spite of failing health wrote accounts of more than one-half of the apple insects and some others, mostly scale insects.

After Professor Slingerland's death in March, 1909, I collected and preserved the manuscript, thinking that it might be possible to publish it without much additional work. While the treatment of each insect was complete in itself, the whole was so disconnected that this plan had to be abandoned. Accordingly, in the summer of 1910 I undertook the task of completing the book, following Professor Slingerland's outline. In doing so, I have made free use of his unpublished notes, and most of the illustrations are from his photographs.

We have attempted to treat only the more important insects injurious to deciduous fruits; many of the minor pests have been omitted altogether. In each case the aim has been to give, in as concise form as possible, the main facts relating to the distribution, life-history, and habits of the insect, the nature and extent of the injury inflicted, and the means of control — the last from the standpoint of the commercial fruit-grower. At the end of the discussion of each insect, references are given to a few of the more important articles relating to the subject. These citations indicate sources of our information and will serve as a guide to those who wish to pursue the subject farther.

I am under great obligations to Professor P. J. Parrott, Dr. O. A. Johannsen, Dr. E. P. Felt, and Dr. Robert Matheson, for carefully reading and correcting the manuscript and for many helpful suggestions. I have been unable to read the proof myself; this tedious but important work has been done by Dr. Matheson, Mrs. Matheson, M. D. Leonard, and by Mrs. Crosby — to them all I give my hearty thanks.

Most of the illustrations are from photographs by Professor Slingerland; a few have been kindly furnished by Professor Herrick, H. H. Knight, and Dr. Matheson. The drawings are by Miss Anna C. Stryke.

C. R. CROSBY.

CORNELL UNIVERSITY, ITHACA, N. Y.,
June 4, 1914.

CONTENTS

CHAPTER I

CHAPTER II

CHAPTER III

CHAPTER IV

CHAPTER IX

CHAPTER X

CHAPTER XI

CHAPTER XII

CHAPTER XIII

MANUAL OF FRUIT INSECTS

MANUAL OF FRUIT INSECTS

CHAPTER I

GENERAL CONSIDERATIONS

INSECTS are among the most formidable enemies to successful fruit-growing. According to conservative estimates from 20 to 40 per cent of the fruit crop of the United States is annually destroyed by insect pests. In 1909 Quaintance estimated the annual loss to the deciduous fruit interests of the country from insect depredations at over $66,000,000, divided as follows:

Codlin-moth	$16,716,667
San José scale	10,528,265
Peach-tree borers	6,000,000
Grape insects	8,769,905
Plum curculio	8,590,769
Miscellaneous apple insects	10,089,932
Miscellaneous pear insects	1,328,613
Cranberry insects	396,656
Miscellaneous stone fruit insects . . .	3,693,843
Total	$66,114,650

This sum includes the cost of spraying and other repressive measures for the control of fruit insects.

The enemies of fruits treated in this book belong, with one or two exceptions, to the class of animals known as insects. The pear leaf blister-mite, the red-spider and clover-mite belong to the Arachnida. In many respects insects differ widely from the higher animals with whose structure we are more familiar.

They have a horny or chitinous external skeleton or shell which serves as a protection to the internal organs and as an attachment for the muscles.   This outer shell is divided into a series of rings or segments.   In insects the body is divided into three well-defined regions, — head, thorax and abdomen (Fig. 1).

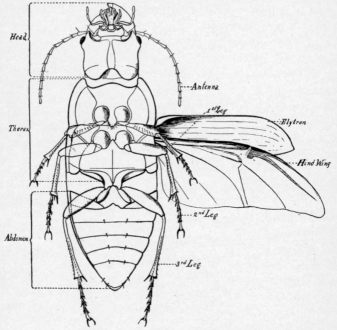

Fig. 1. — Ventral view of a beetle.

The head bears a pair of compound eyes, two or three simple eyes, a pair of antennæ and the mouth parts.   The thorax is composed of three segments and bears on the inner side three pairs of legs, one pair to each segment, and on the upper side two pairs of wings, a pair on each of the last two segments. In the flies, only one pair of wings is present, the hind pair being represented by a pair of knobbed appendages known as poisers. The wings are variously modified for different uses. In the beetles the front pair are very hard and horny and not suited for flight, but fitting closely together serve as a protection to the hind wings which in repose are folded under them (Fig. 1).

The abdomen consists of from ten to twelve segments. In many species the tip in the female is provided with a sharp lancelike or saw-edged oviposi-tor, with which she punctures the tissue of plants and in-serts her eggs in the wound so made (see Fig. 317, p. 358).

FIG. 2. — A caterpillar feeding, showing the biting type of mouth parts.

How insects feed.

The mouth parts of insects are adapted for feeding on all sorts of tissue, from the tender leaves and ripening fruit to the solid wood itself. From the standpoint of control it is of great importance to know just how each insect obtains its food.

For the purpose of control insects may be roughly divided into three classes as follows:

1. *Chewing insects:* Beetles and caterpillars belong here. They are provided with hard horny jaws or mandibles with which they bite off and swallow portions of the tissue of plants as shown in Figure 2. It is usually possible to kill such insects by poisoning their food with an arsenical.

2. *Sucking insects:* Plant lice and other true bugs are furnished with a beak containing four bristles united into a

slender tube.   In feeding, the tip of the beak is applied to the
surface of the plant, the bristles are inserted into the tissue and
the plant juices are sucked
out (Fig. 3).   Contact insecticides must be used against
this class.

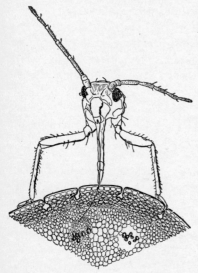

3. *Lapping insects:* In the
fruit flies the mouth parts
are developed into a tongue-
like organ with which the
insect is able to lap or lick
up liquids (Fig. 4). Arsenical
poisons have been used successfully for the control of
this class of insects.

In different stages of its
development the same insect
may have different kinds of
mouth parts, and may feed
on entirely different foods;
for instance, caterpillars have
biting mouth parts and may

FIG. 3. — A plant-louse feeding, showing the sucking type of mouth parts.
From a German drawing.

feed on leaves, while the adults, moths,
have sucking mouth parts with which
they extract the nectar from flowers.

How insects breathe.

Insects do not possess lungs, but
breathe through a series of openings
called spiracles extending along each side
of the body.   These openings connect
with tubes called tracheæ, which, sub-
dividing again and again, extend to all
parts of the body.   Some contact insecticides are supposed to clog these

FIG. 4. — Head of a
fruit-fly, showing the lapping type of mouth parts.

tubes and so smother the insect, while others, like the oils, are said to penetrate the thin walls of the tracheæ and thus reach a vital part in the internal organs of the insect.

The development of insects.

Most insects, with the exception of some scale insects and certain forms of plant lice, reproduce by means of eggs. The newly hatched insect usually bears little resemblance to the adult. As it increases in size its skin becomes too small and a new skin is formed beneath the old one and the latter is discarded; this is known as molting. The period between two successive molts is called an instar. The number of instars varies in different insects from three to six or seven; five is the more common number. In some insects the change from the immature condition to the winged adult takes place without any material change in form; in others the transformation is abrupt and striking. In the former case the insect is said to have an *incomplete metamorphosis;* in the latter a *complete metamorphosis.*

Incomplete metamorphosis.

In this type of development the immature stages resemble the adult in form. The wings develop externally as pad-like outgrowths of the thorax but do not become functional till the adult stage is reached. The immature forms are known as *nymphs.* In this type of development the life cycle of the insect consists of three stages, viz. the egg, the nymph (3–5 instars) and the adult. The true bugs and grasshoppers have incomplete metamorphosis.

Complete metamorphosis.

In this case the immature stages of the insect bear little or no resemblance to the adult. The wings develop internally in pockets formed by an infolding of the body wall of the thorax. The immature stages are known as larvæ. The larva molts five or six times, and when full-grown transforms to an inactive pupa, usually in a cocoon or earthen cell prepared by the larva. The pupa is a resting stage in which the organs of the larva are broken down and made over into those of the adult. In the pupa the antennæ, legs and wings of the adult are usually evident, closely applied to the body and covered by the pupal skin. When the remarkable internal structural changes in the pupa are complete the adult winged insect emerges. In this type of development there are four stages, viz. egg, larva (5–6 instars), pupa, adult. Butterflies, moths (caterpillars) and beetles have complete metamorphosis.

The control of insects.

In spite of the many natural checks to which insects are subject, such as extremes of temperature, drought and wet, the depredation of parasitic and predacious enemies, and the results of fungous and bacterial diseases, it is usually necessary to protect the fruit crop by artificial means. The control of each insect is a special problem, and its solution requires a thorough knowledge of the life history and habits of the insect as well as an understanding of the nature of the crop and the conditions under which it is grown.

Substances used for killing insects are known as insecticides, and are discussed in the last chapter of this book, page 474.

Spraying.

Insecticides are most commonly applied in the form of a liquid by means of a spray pump. In the case of poisons the object is to cover the foliage evenly with the spray so that the insect will be sure to get it in feeding; in the case of contact insecticides it is necessary to hit each insect. Spraying is an art, and the finer points can only be learned by practice. Success will depend on the timeliness of the application and the thoroughness with which the work is done. Young insects are killed more easily than old ones, and it is easier to kill a few early in the season than to wait until they have become abundant before spraying. In spraying the grower should always have a definite object in view. He should study his trees, find out their needs, and treat them accordingly. A general spray given with a vague hope of " doing some good " rarely ever pays. It is important that the application be made at the proper time. Development of the buds and blossoms should be watched and the spraying timed accordingly. It is not safe to go by the calendar. Unless the spraying is done thoroughly it is likely to be wasted effort — do a good job but do not overdo it. Use enough liquid and put it in the right place at the right time, thus avoiding waste and unnecessary expense. A sprayer should be selected adapted to the crop and to the size of the orchard. Use the best nozzles, pump and engine you can buy; a good workman deserves good tools, and they pay in the long run. Use only standard spray materials and buy them of reputable dealers. It is expensive business to experiment with quack remedies or to use cure-alls of unknown composition.

Dusting.

Insecticides are sometimes applied in the form of a dust. The results obtained by dusting have in general not been so satisfactory as those obtained by spraying, but more recent experiments where finer materials were used have shown that this method may be of great value. It is especially useful where the supply of water is not convenient or limited.

Clean farming.

In many cases the application of insecticides alone is not sufficient to control orchard insects, but must be supplemented by other measures. The accumulation of dead leaves, grass and weeds along fences or in hedgerows provides ideal winter quarters for many insects. Stone walls, stone piles, and similar shelters make the control of the plum curculio unnecessarily difficult. Uncultivated apple orchards are usually more susceptible to injury by curculio, apple maggot, leaf miners, and insects with similar hibernating habits. By practicing clean farming many of these pests will be reduced to a minimum, so as to be more readily controlled by spraying. Furthermore, trees that are in a vigorous, healthy condition from proper cultivation are as a rule less subject to spray injury and are, therefore, able to receive stronger and more frequent applications of insecticides without danger.

Crop rotation.

In the case of strawberries and some other small fruits a frequent rotation of crops will tend to clear the land of many insect pests which cannot be reached satisfactorily in other ways. This is particularly true of wire-worms, white grubs and the strawberry root-worm.

CHAPTER II

APPLE INSECTS — THE FRUIT

Nearly five hundred species of insects have been recorded as feeding on the apple, but fortunately, the greater part of them do not cause enough damage to be considered of economic importance and are not treated in this book. Many of the most important apple insects have come to us from foreign lands; the codlin-moth and bud-moth from Europe, and the San José scale from China; others fed originally on the wild thorn, as the apple maggot and apple curculio, but have found in our orchards an abundance of food and other conditions better to their liking. Many apple insects also attack the pear and quince.

In spraying apples the insecticide can usually be combined to advantage with a fungicide for the control of apple scab. The following spraying schedule for the San José scale, codlin-moth, bud-moth, case-bearers, and apple scab is intended for New York State conditions, but with modifications might be used in other regions.

Dormant spray.

As the leaf buds begin to show green

Lime-sulfur (32 degrees Beaumé), diluted 1 to 8, for San José scale, oyster-shell scale and blister-mite; add two pounds arsenate of lead to 50 gallons of mixture for bud-moth.

Summer sprays.

A. — As blossom buds begin to show pink

Lime-sulfur (32 degrees Beaumé), diluted 1 to 40, for apple scab. Add arsenate of lead, 2 pounds to 50 gallons, for bud-moth and case-bearers.

9

B. — *As the last of the petals are falling*

Lime-sulfur (32 degrees Beaumé), diluted 1 to 40, for apple scab. Add arsenate of lead, two pounds to 50 gallons, for codlin-moth.

This is the most important spray for the control of the codlin-moth and should be thoroughly done.

C. — *Three weeks after the petals fall*

Lime-sulfur (32 degrees Beaumé), diluted 1 to 40 for apple scab. Add arsenate of lead, two pounds to 50 gallons, for codlin-moth.

D. — *Last week in July*

Lime-sulfur (32 degrees Beaumé), diluted 1 to 40 for apple scab. Add arsenate of lead, 2 pounds to 50 gallons, for second brood of codlin-moth.

The Codlin-Moth

Carpocapsa pomonella Linnæus

This is by all odds the most destructive insect enemy of the apple. Originally a native of southeastern Europe, it has now become nearly cosmopolitan, occurring in all the apple-growing regions of the world. It was introduced into New England some time before 1750, and spread gradually westward, reaching Iowa about 1860, Utah in 1870 and California about 1874.

The amount of injury to the crop varies greatly with the climate. In the Northern states and Canada the injury in unprotected orchards averages from 25 to 50 per cent of the crop, while in the South and in the warmer valleys of some of the far Western states losses of from 60 to 95 per cent are not uncommon. This greater destructiveness of the moth in the South is the result of the longer growing season, which permits more generations to develop than are possible in the North. The larvæ of later generations are much more numerous than the first, and the injury consequently greater. Quaintance in 1909 estimated the annual loss to the fruit industry of the United States from this insect alone at over $16,000,000, three fourths of this being direct injury to the crop and the other one fourth the cost of spraying and spray materials.

Almost invariably the codlin-moth hibernates as a larva in tough silken cocoons under loose pieces of bark, in crevices in the tree or in near-by fences, or in other suitable shelter (Fig. 5). In orchards of smooth-barked trees where no better place is available they will spin their cocoons in cracks in the ground and at the base of the trunk. The cocoon is rather thin, but quite tough, and is made largely of silk in which are mixed bits of the substance on which it is made.

Fig. 5. — Codlin-moth larva in its cocoon on a bark-flake (× 2).

It is lined with white silk, and the outside is rendered quite inconspicuous by the addition of bits of dirt and bark. The cocoons of the over-wintering larvæ are thicker and tougher than those of the summer broods.

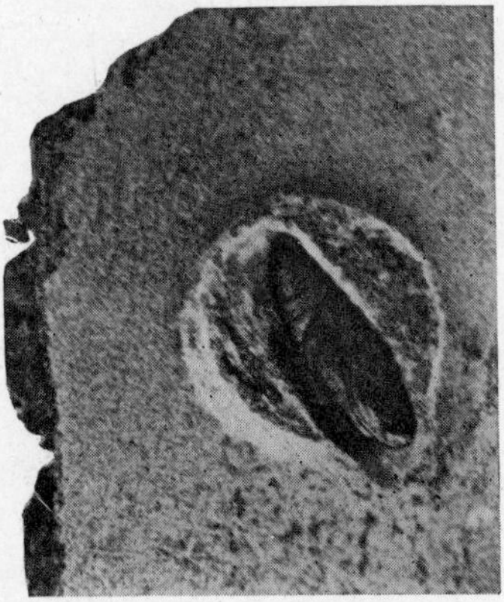

With the advent of warm weather in the spring the larva changes to a pupa, usually within the old cocoon (Fig. 6), but sometimes it may migrate to some other portion of the trunk and there construct a new cocoon. Usually just before pupation the larva will open the

Fig. 6. — Codlin-moth pupa in cocoon (× 2).

end of the cocoon and spin a tube of silk out to the surface, stretching a thin sheath of silk across the opening at the ending of the cocoon, which is ruptured when the moth emerges. The time spent in the pupal state varies with the climate, but

averages not far from 26 or 28 days for the spring brood. The pupa is about $\frac{1}{2}$ inch in length and varies according to age from yellowish to brown. Its back is armed with transverse rows of minute spines, by means of which it is able to push itself part way out of the cocoon just before the moth emerges.

Fig. 7. — Codlin-moth, adult. Knight photo ($\times$ 3).

The adult is a beautiful moth about $\frac{3}{4}$ inch across the expanded wings (Figs. 7 and 8). The front wings have the general appearance of watered silk, this effect being produced by alternating irregular lines of brown and bluish gray. Near the hind angle is a large, light brown area bounded on the inner side by an irregular chocolate brown band and crossed by two similar bands of a metallic coppery or golden color in certain lights. The hind

Fig. 8. — Codlin-moth resting on a small apple. Knight photo ($\times$ 2$\frac{1}{2}$).

wings are coppery brown, darker towards the margin. The sexes are very similar, but the male may be distinguished by the presence of an elongate dark area on the underside of the fore wing and a pencil of black hairs on the upper surface of the hind wing. The moths continue to emerge for a period of several weeks, but the majority appear about a week after the petals fall from the apple blossoms, the exact time depending on the character of the season. If the weather is warm, egg-laying begins in from 3 to 5 days, but if it is cold, the moths may remain inactive for a long period. The average life of the moth is about 10 days, and each female lays from 30 to over 100 eggs.

Fig. 9. — Two views of the codlin-moth egg-shell ($\times$ 22 and $\times$ 7).

The glistening, flat, oval, scale-like eggs (Fig. 9) of the spring generation are laid mostly on the upper and under surfaces of the leaves, although a few may be found on the fruit and branches. The egg is about $\frac{1}{25}$ inch in diameter,

Fig. 10. — Newly hatched codlin-moth larva ($\times$ 44).

or about half the size of a pinhead. The egg-laying period extends over several weeks, as is the case with the emergence of the moths, but the majority are laid about two weeks after the falling of the petals. The spring brood eggs hatch in 6 to 10 days, while the eggs of the later generations, because of the higher temperature, require a much shorter period, 5 or 6 days.

The time at which the first brood eggs hatch and the young larvæ enter the fruit is of great importance from the standpoint of control, and has been given careful attention by entomologists. In spite of many disturbing factors it may be said that in general the majority of the eggs hatch about 3 or 4 weeks after the petals fall. The newly hatched larvæ (Fig. 10) are less than $\frac{1}{16}$ inch in length and are of a semitransparent whitish color, with a shiny black head and blackish thoracic and anal shields. At first they may feed slightly on the foliage, but usually they crawl directly to the fruit, which they enter in the majority of cases at the blossom end. A few enter at the stem end and a considerable number commence their burrows where a leaf or another apple touches the surface of the fruit. In the case of later generations a much greater proportion of the larvæ enter at the side, a fact of great practical importance. After feeding slightly in the calyx cup the larva burrows directly to the core, where it devours the seeds and eats out a considerable cavity, leaving it partially filled with a filthy mass of excrement loosely webbed together with silk (Fig. 12). Larvæ of later generations entering at the side frequently eat out a small burrow or cavity just beneath the skin before starting the burrow towards the core.

Fig. 11. — Full-grown codlin-moth larva (× 2½).

When nearly full grown the larva eats out a new burrow directly to the surface but keeps the opening plugged with excrement until it is ready to leave the fruit (Fig. 14). The time spent in the fruit varies considerably, but averages not far from

30 days for the first brood and 3 weeks for the second. The full grown larva is about ¾ inch in length, pinkish white in color, with the head dark brown and the thoracic and anal shields lighter brown (Fig. 11).

The larger part of the larvæ leave the fruit before it falls and crawl down the branches until they find a suitable place for spinning the cocoon. After making the cocoon the larva may do one of two things, either it will remain in the larval condition until the following spring or it may change to a pupa in about a week. In the latter case these summer pupæ give rise in about 10 days to a brood of moths which lay the eggs for the second generation. In New York only part of the larvæ spinning up before August 1 transform the same season, all the larvæ going into cocoons after that date hibernate. In Arkansas the corresponding date is September 1. Not all the larvæ of the first broods transform the same season even in Georgia, where there are three full generations, but they are relatively few.

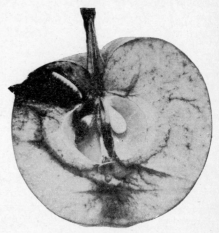

Fig. 12. — Codlin-moth larva in its burrow in an apple.

The number of generations a year varies in different parts of the country, and has been a difficult question to solve, owing to the overlapping of the broods. The earliest of the first brood moths will be on the wing before belated individuals of the spring brood have disappeared. It is now, however, pretty well determined that in the North, — New England, New

York and Michigan, — there is one full generation and usually a partial second, the completeness of the latter depending on the length of the season. In Nebraska, Missouri and Virginia there are two full generations; in Arkansas there are three and in Georgia three and a partial fourth. In Washington, Oregon, Utah and Idaho there are two full generations and there is strong evidence that there are at least three in Arizona.

FIG. 13. — A wormy apple, showing the mass of brown particles thrown out at the blossom end by the young larva.

While the codlin-moth is distinctly an apple pest, it is also an important enemy of the pear. In 1898 Slingerland estimated the loss to the pear crop in New York at $500,000. Wild haws, crab apples and quinces are also quite freely eaten by the larvæ. In California English walnuts are generally infested to a slight extent by the larvæ of the later generations. It has been reported as injuring plums in Canada and also in New Mexico.

Natural enemies.

The eggs of the codlin-moth are parasitized by a minute chalcis-fly, *Trichogramma pretiosa* Riley, four of these tiny flies having been reared from a single egg. The eggs are also attacked by a mite, *Trombidium* sp. The larva is attacked by the hymenopterous parasites, *Pimpla annulipes* Brullé, *Macrocentrus delicatus* Cress., *Ascogaster carpocapsæ* Vier., *Goniozus* sp., *Bethylus* sp., and by two Tachina flies, *Hypos-*

tena variabilis Coq. and *Tachinophyto* sp. In Georgia a small chalcis-fly (*Haltichella* sp.) has frequently been reared from the pupa. A European parasite (*Calliephialtes messor* Grav.) has been introduced into California, but apparently with little success. The larvæ of a num-

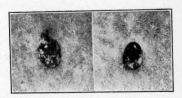

ber of beetles have been found killing the larvæ and in the South several species of ants attack the larvæ and pupæ in the cocoons. In Utah a wasp stocks its burrows with the larvæ. Both in this country and in Europe larvæ have been found infested by hair-snakes.

FIG. 14. — Exit hole of codlin-moth larva; left, before leaving the fruit; right, after it has emerged and pushed away the plug.

In spite of this array of insect enemies the codlin-moth is able to maintain itself as the most destructive enemy of apples

FIG. 15. — Empty codlin-moth cocoons on the under side of a flake of bark; view of the outer surface of the same flake showing the holes made by birds in reaching the larvæ.

and pears. Its most effective natural enemies are the birds, over a dozen species of which are known to feed on it. The downy woodpecker, nuthatch and chickadee destroy great numbers of the hibernating larvæ, under loose flakes of bark. In fact, it requires diligent search to find larvæ towards spring even where empty cocoons are abundant. Usually a tell-tale hole through the bark flake into the cocoon explains the absence of its occupant (Fig. 15). These birds are such efficient aids to man in controlling the codlin-moth that they should

Fig. 16. — Apple and pear fruits with the calyx lobes still expanded; the right time to make the first spraying for the codlin-moth.

be carefully protected. During the winter they feed in small flocks, going over the same territory day after day, carefully examining every portion of the bark for insect food. They may be induced to visit an orchard regularly by tying strips of beef fat to a few of the branches and the destruction of codlin-moth larvæ will more than pay for the trouble involved.

Means of control.

When single brooded or when the second generation is only partial, the codlin-moth has not been found a very difficult

pest to control by spraying with an arsenical poison. In
the South and portions of the West, however, where two
or more full generations develop, spraying has in general
given less satisfactory results. In spraying for this insect
advantage is taken of the fact that the great majority of
the young larvæ enter the apple at the blossom end. For
about two weeks after the petals fall the calyx lobes are spread
wide apart and the young apple stands upright on the stem

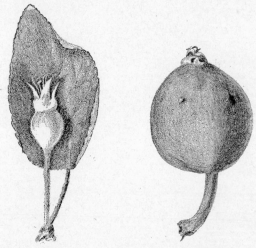

Fig. 17. — Two later stages, the calyx lobes closed; it is now too late to spray
effectively.

with the calyx end directed upward (Fig. 16). If at this time
an arsenical spray is thoroughly applied with sufficient force
to drive the poison into the calyx cup, minute particles of the
poison will be deposited where the young larva will get it in
his first meal as it enters the fruit. In about two weeks the
calyx lobes close as shown in Figure 17, and it is then too late
to spray effectively, for it is then impossible to place the poison
where it will do the most good. The closed calyx lobes form

a tight roof over the cavity and prevent the rain from washing away the poison. To be most effective this first spray should be applied as soon as possible after the larger part of the petals have fallen. Great care should be taken to hit each apple, the spray should be directed downward directly into the blossom end of the fruit and sufficient power should be used to give a strong spray. In commercial orchards the best results are obtained where a good power sprayer is used and where the nozzle, on the end of a light extension rod, is handled by a man standing on an elevated platform or tower. In the Eastern states a fine, mist-like spray is most commonly used, but in the Far West remarkable results have been obtained by the use of a coarse driving spray, such as is produced by the Bordeaux nozzle.

Not all the larvæ are killed in the calyx cavity, for quite a number always enter the fruit at some other point. A large part of these are killed by the poison on the leaves where many of them feed slightly before reaching the fruit, while others are doubtless destroyed by the poison adhering to the surface of the apple, although the number killed in this way is not large.

This first spraying immediately after the petals fall is the most important operation in the fight against the codlin-moth, and no pains should be spared to make it as effective as possible. Not only does it control the injury by the first brood larvæ, but it also prevents in large measure the losses occasioned by the later broods. In the Far West, where two full broods develop, some remarkable results have been obtained from this spraying alone when the application was made with great thoroughness, using a coarse driving spray and sufficient pressure to place the poison deep in the calyx cavity. In some cases as high as 95 or 99 per cent of the crop has been protected in this way without the necessity for any later spraying. Attempts to control the codlin-moth in the East by the one-spray method have not as yet shown it to be superior to the more common practice under Eastern conditions, where it is necessary to make repeated applications of a fungicide for the control of apple scab and other fungous diseases, but they have called attention to the great importance of doing very thorough work with the first spray.

At the time the first spraying is made the codlin-moth eggs have not yet been laid and the majority do not hatch until about 3 or 4 weeks later. If about the time of hatching the foliage and fruit are thoroughly coated with a fine arsenical spray, many of the newly hatched larvæ will be killed before reaching the apple, since many of the eggs are laid at some distance from the fruit and the larvæ feed to some extent on the leaves.

In case the first spraying, because of carelessness or for some other reason, has not controlled the worms, it may be advisable to spray for the second brood. This spraying should be done just as the majority of the eggs are hatching. The proper time may be determined by banding a few trees with burlap bands. When empty cocoons are found beneath them it shows that the moths are emerging. Eggs will be hatching in about a week or two. In New York the second brood larvæ enter the fruit in late July and in August, but the exact time varies greatly with the season.

The larvæ of the later broods are much more numerous than those of the first and the loss which they inflict is correspondingly greater. Where the first spraying has been neglected one cannot hope to protect his crop by spraying for the second brood alone. But where the first brood has been reduced to a minimum by a thorough early spraying much good can oftentimes be accomplished by a later spraying to destroy the progeny of the few stragglers missed earlier in the season.

Paris green has been for years the standard poison used against the codlin-moth, but it has now been almost entirely replaced by arsenate of lead. The latter has, on the whole, given better results. It sticks better to the foliage and fruit, contains practically no free arsenic and may be combined with the dilute lime-sulfur, as used for the apple scab. One pound of Paris green or 4 to 6 pounds of arsenate of lead in 100 gallons of water has, in general, given the best results. Paris green or arsenate of lead may be combined with Bordeaux mixture and arsenate of lead with lime-sulfur, but Paris green has proved injurious to the foliage when used with the latter. Success in controlling the codlin-moth does not depend so much on the kind of poison used as on the thoroughness and timeliness of the application. The personal factor is of the greatest importance.

At picking time many infested apples are carried to the packing shed or storehouse, where the larvæ emerge and spin up in cracks and crevices. In case the storehouse is near the orchard the windows should be screened to prevent the escape of the moths the following spring.

Before the discovery of the arsenical method of controlling the codlin-moth banding the trunks with strips of burlap was widely practiced. These bands are put around the trunk and larger branches, and beneath them a large proportion of the larvæ will spin their cocoons. They should be examined and the larvæ killed regularly every week during the cocooning season. It is generally believed that in properly sprayed orchards the use of bands will not pay for the trouble and expense involved.

REFERENCES

Howard, Rept. U. S. Dept. Agr. for 1887, pp. 88–115.
Cornell Agr. Exp. Sta. Bull. 142. 1898. *Extensive bibliography.*
U. S. Bur. Ent. Bull. 41. 1903.
Utah Agr. Exp. Sta. Bull. 87. 1904.
Utah Agr. Exp. Sta. Bull. 95. 1906.
Wash. Agr. Exp. Sta. Bull. 77. 1906.
N. M. Agr. Exp. Sta. Bull. 65. 1907.
Ill. Agr. Exp. Sta. Bull. 114. 1907.

Quaintance, Year Book U. S. Dept. Agr. for 1907, pp. 435–450.   1908.
N. H. Agr. Exp. Sta., 19th and 20th Rept., pp. 396–498.   1908.
U. S. Bur. Ent. Bull. 80, Pt. I.   1909.
Mo. State Fruit Exp. Sta. Bull. 21.   1909.
Ga. State Bd. Ent. Bull. 29.   1909.
Felt, 25th Rept. N. Y. St. Ent. pp. 25–71.   1910.   *Extensive bibliog-
    raphy.*
U. S. Bur. Ent. Bull. 80, Pt. V.   1910.
U. S. Bur. Ent. Bull. 80, Pt. VI.   1910.
Ont. Dept. Agr. Bull. 187.   1911.
Wash. Agr. Exp. Sta. Bull. 103.   1911.
U. S. Bur. Ent. Bull. 97, Pt. II.   1911.
Felt, Jour. Ec. Ent. V, pp. 153–159.   1912.
U. S. Bur. Ent. Bull. 115, Pts. I, II.   1912.

THE LESSER APPLE WORM

Enarmonia prunivora Walsh

This insect is closely related to the codlin-moth, and both in its life history and the nature of its injury to the apple is very similar to that insect.   Originally described by Walsh in 1867 as a plum pest, it has more recently shown itself a serious enemy of the apple in certain localities. It is generally distributed throughout the Eastern states from Texas and Georgia northward to Ontario and Quebec and also occurs in British Columbia.

The full-grown larvæ are about $\frac{3}{8}$

FIG. 18. — Upper figure, partly grown codlin-moth larva; lower two, larvæ of the lesser apple worm (× 6).

FIG. 19. — Cocoon of the lesser apple worm with empty pupa skin protruding (× 7).

inch in length, and pinkish or nearly white in color (Fig. 18). They leave the fruit in the fall a little later than the codlin-moth larvæ and spin cocoons in similar situations; occasionally pupation occurs within the apple itself (Fig. 20). The larva may be distinguished from that of the codlin-moth by its smaller size, and by having a brownish, comb-like structure on the caudal curvature of the anal plate visible only under a strong lens.

The cocoon is about ¼ inch long, lined with white silk and covered on the outside with bits of bark and dirt (Fig. 19). The following spring the larva changes to a brownish pupa less than ¼ inch in length. The dorsal surface is armed with transverse rows of short spines by which the pupa works itself

FIG. 20. — Empty pupa skin of lesser worm protruding from end of small apple.

part way out of the cocoon before the emergence of the moth. The empty pupa case is usually left attached to the cocoon.

The moth (Fig. 21) measures about $\frac{7}{16}$ inch across the expanded

FIG. 21. — Lesser apple worm moth (× 5).

wings. The front wings have the general appearance of watered silk when viewed with the unaided eye. The general color is a warm brown, darker on the front margin and at the tip and lighter towards the base. Under a lens it is seen that this effect is produced by groups of scales of three shades of brown; reddish, chocolate and light brown; in addition the wing is crossed by three fine interrupted irregular pearl-blue lines, the basal one being double in front. The hind wings are brown, paler towards the base.

The moths emerge and eggs are laid at about the same time as those of the codlin-moth. The eggs are glistening milky white, flat and scale-like, and closely resemble those of the codlin-moth, but are only a little more than one half as large. They hatch in 5 or 6 days.

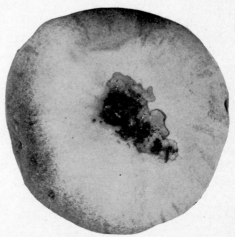

FIG. 22. — Work of lesser apple worm at the blossom end of a mature apple.

The injury caused by this insect has often been confused with the work of the codlin-moth. Many of the young larvæ enter the fruit through the calyx cavity, while others bore through the skin just outside the calyx, and still others enter at the side and near the stem. A larger proportion of the first brood enter at the calyx than is the case with the second generation. In general, the burrows are not so deep as those of the codlin-moth. The larva remains near the surface and eats out a blotched mine just beneath the skin which turns whitish and greatly disfigures the fruit (Fig. 22).

In the North there are two generations annually; in the Ozark region there are three and sometimes a partial fourth brood.

Many of the first brood larvæ pupate in the fruit and the empty pupa case is frequently found protruding from the burrow. Apples infested by small larvæ are frequently placed in storage where they continue to feed and often cause considerable loss.

The lesser apple worm moth has also been reared from plum, from black-knot, a fungous swelling on plum branches, and from certain insect galls on elm and oak.

Remedial measures.

The treatment suggested for this insect is the same as that for the codlin-moth, except that there is especial need of making the second spraying, 3 to 4 weeks after the petals fall, very thorough. To kill the young larvæ entering at the stem and side at that time the foliage and fruit should be thoroughly coated with a fine arsenical spray.

REFERENCES

U. S. Bur. Ent. Bull. 68, Pt. V. 1908.
Taylor, Jour. Ec. Ent., II, pp. 237–239. 1909.
U. S. Bur. Ent. Bull. 80, Pt. III. 1909.

Apple Fruit-miner

Argyresthia conjugella Zeller

The larva of this small Tineid moth is a serious pest in the apple orchards of western Canada. It also occurs in northern Europe, where it frequently destroys the entire apple crop. In Europe it originally fed on the berries of the Mountain Ash and in Canada on the fruit of the Wild Crab (*Pyrus fusca*) but it has now become thoroughly established on the cultivated apple. In England and Scandinavia it has been found infesting the cherry.

The injury is caused by the pinkish white larva, about $\frac{3}{8}$ inch in length, which burrows in all directions through the fruit during July, August and September. The tissue around the burrows turns brown, decay ensues and the apple is ruined. The parent moth (Fig. 23) has a spread of nearly $\frac{1}{2}$ inch; the front wings are iridescent purplish gray mottled with brownish; on the front margin is a row of minute white and brown dots and a larger oblique white mark occurs near the apex; on the hind margin is a broad creamy-white band interrupted near the middle by a brownish spot. The moths appear in May and June. The eggs are unknown.

The young larvæ enter the fruit at the side and on becoming full-grown leave the apple and seek shelter under the bark on the trunk or under leaves on the ground. The winter is passed

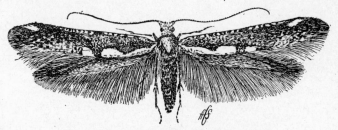

Fig. 23. — The apple fruit-miner moth ($\times 7\frac{1}{2}$).

in the pupal state in white cocoons, the outer layers of which are loose and have the threads arranged so as to form a beautiful openwork pattern.

Remedial measures.

Satisfactory methods of control have not yet been devised, but several thorough sprayings with arsenate of lead, so applied as to keep the fruit coated with the poison, would doubtless do much to lessen the injury.

References

Rept. Exp. Farms Ottawa 1896, pp. 258–262, 1897.
Rept. Exp. Farms Ottawa 1897, pp. 201–202, 1898.
Reh, Prakt. Ratg. Obst- und Gartenbau, XXII, pp. 452, 453. 1907.

The Apple Red Bugs

Heterocordylus malinus Reuter, and *Lygidea mendax* Reuter

These two native sucking plant-bugs have in recent years caused considerable injury in certain orchards in New York and New Jersey by puncturing the young apples during May and early June. Many of the punctured apples fall to the ground, others dry up on the tree, while the remainder mature but are badly deformed and rendered unmarketable (Figs. 24 and 25). Red bug injury may be distinguished from the work of the plum curculio by the fact that in making the puncture the insect does not remove any of the

Fig. 24. — Mature apple deformed by red bug punctures.

tissue, but merely sucks out the juices. When abundant the apple aphis frequently causes knotty and misshapen apples, but its work is usually characterized by a stunting or pucker-

Fig. 25. — Young apples which dropped prematurely as a result of red bug injury.

ing of the blossom
end which is not
present in typical
red bug injury.

The life histories
of the two species
are very similar.
The dull whitish,
strongly curved,
slightly compressed
eggs are inserted
their full length
into the bark on the
smaller branches
(Figs. 27 and 28).

FIG. 26. — Red bug nymph feeding on a newly set apple. Much enlarged.

They hatch soon after the opening of the leaves of the fruit
buds and the minute, tomato-red nymphs at once begin to
puncture the tender
leaves. The clusters of
minute reddish dots
caused by these punc-
tures are quite conspicu-
ous and are usually the
first indication of the
presence of the nymphs.
The injury to the foliage
is very slight. They
may feed on the leaves
until full grown but usu-
ally attack the fruit as soon as it sets (Fig. 26).
In the case of very small apples, the four sharp
bristles of the beak penetrate quite to the center,
the surrounding tissue becomes discolored and hardened and
the apple is ruined.

FIG. 27. — Eggs of *H. malinus* inserted in a slit in the
bark at the base of a fruit spur.

FIG. 28. —
Eggs of *L.
mendax* in len-
ticels on a
two-year-old
apple branch.

The young nymphs of the two species are very similar. Those of *L. mendax* may be distinguished by their brighter red color,

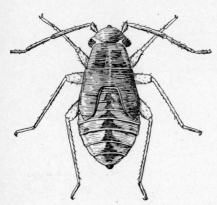

the absence of dusky markings on the thorax and by having the body clothed with fine short black hairs. Both species pass through five immature stages and attain wings at the fifth molt (Figs. 29 and 30). The adults of both species are about $\frac{1}{4}$ inch in length. In *H. malinus* (Fig. 31) the general color varies from red to nearly black and the entire dorsal sur-

FIG. 29. — Fifth stage nymph of *H. malinus* (× 9).

face is sparsely clothed with conspicuous white, flattened, scale-like hairs. In *L. mendax* (Fig. 32) the general color is lighter and these hairs are lacking.

As far as we have observed Greenings, Pound Sweets and Spies, in the order named, are the varieties most subject to attack. Sometimes the whole crop is rendered unmarketable, but such severe injury is unusual.

Remedial measures.

It has been found impracticable to attempt to destroy either the eggs or the adults. The former

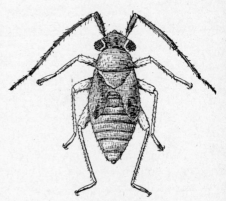

FIG. 30. — Fifth stage nymph of *L. mendax* (× 9).

are inserted in the bark where the embryo develops some distance from the surface; the latter are relatively few and

Fig. 31. — *H. malinus*, adult (× 3¾).

occur when the trees are in full foliage, when it would be very difficult to hit them. Attempts to destroy the young nymphs by spraying with kerosene emulsion or whale-oil soap have not been successful, but fairly good results have been obtained by the

Fig. 32. — *L. mendax*, adult (× 3¾).

use of "Black Leaf 40" tobacco extract, one pint in 100 gallons of water, applied very thoroughly just before the blossoms open. Sometimes a second application, just after the falling of the petals, may be found necessary. "Black Leaf 40" can be used with the lime-sulfur as used for a summer spray. When used with water add 4 to 5 pounds of soap to make the mixture stick and spread better. The spraying should be done on bright warm days, for in cool weather many of the nymphs hide away in the opening leaves.

REFERENCE

Cornell Agr. Exp. Sta. Bull. 291. 1911.

THE APPLE MAGGOT

Rhagoletis pomonella Walsh

This native American insect, although originally feeding in the fruit of the wild thorn, has during the past sixty years become a serious enemy of the apple in the Eastern states and

Canada.   While summer and early fall varieties are particularly
subject to attack, winter apples are also sometimes badly in-

FIG. 33. — Full-grown apple maggot, side view (× 7).

fested.   Sweet and
subacid varieties are
most susceptible, but
such acid varieties as
Greening, Baldwin
and Oldenburg are
sometimes attacked.
In the Lake Cham-
plain region the Fameuse is very subject to injury, and in
western New York and Canada crab apples are sometimes
badly infested.

The injury is caused
by a whitish maggot,
$\frac{1}{4}$ inch or more in
length, which bur-
rows in all directions
through the fruit
(Figs. 33 and 34).

FIG. 34. — Full-grown apple maggot, ventral view.

In the Northern states, the parent flies appear in early July
and continue abundant well into September.   The females do

FIG. 35. — Apple maggot fly (× 4¾).

not begin egg-laying till
two or three weeks after
emergence. During this
time they may be seen
resting on the leaves or
fruit and lapping up
drops of moisture, or
licking the surface of the
waxy covering of the
fruit with their fleshy

proboscis.   They are blackish, two-winged flies with the head and
legs yellowish ; the abdomen has three or four transverse white

bands and the wings are crossed by four dark confluent bands
(Fig. 35). They are slightly smaller than the house fly, which
they closely resemble in shape (Fig. 37). The female is provided with a sharp ovipositor with
which she punctures the skin of the
apple, usually on the side, and inserts her minute, whitish, elongate
egg directly into the pulp. The eggs
hatch in from two to six days. On
hatching the young maggots start
their tunnels through the flesh but
grow very slowly until the fruit begins to ripen or soften from decay

FIG. 36. — Apple maggot puparia (× 7).

(Fig. 38). When this occurs the maggots grow rapidly, and
by their winding burrows soon reduce the interior to a brownish,
sponge-like mass. It frequently happens that at picking time

FIG. 37. — Apple maggot fly resting
on an apple (× 2).

the fruit may show no signs of
infestation, only to go down suddenly from maggot attack after
having softened in storage or in
transit. This is very likely to
happen in the case of Fameuse
and Mackintosh when grown in
infested localities. Sometimes
the burrows run for some distance just beneath the skin,
showing through as darkened
trails, from which the insect has
received, in some localities, the
name of railroad worm. When
full-grown, the larva escapes
through a ragged opening in the skin of the fruit, usually after
it has fallen, and then as a rule burrows an inch or so into
the soil, where it hibernates in a brownish puparium (Fig. 36),

D

which has been aptly likened to a grain of wheat. In New York, at least, there is a partial second brood of flies appearing in September.

Remedial measures.

As the eggs are inserted directly into the pulp beneath the skin of the fruit, and as the maggots never leave the apple until full-grown, it is impossible to kill them

Fig. 38. — Apples infested with apple maggot beginning to decay.

with any poison or contact spray. The flies, however, can be readily destroyed by having the fruit and leaves covered with an arsenate of lead spray at the time of their emergence in early July. As stated above, the flies do not begin oviposition until three or four weeks after emergence; during this time they feed considerably on the waxy covering of the fruit and lap up drops of moisture from the fruit and foliage. Experiments in New York have shown that if the trees are sprayed the first week in

July with arsenate of lead, 4 pounds in 100 gallons of water, most
of the flies will be killed. It has been suggested that the addition of molasses or sirup to the poison spray would make it more
attractive to the flies, but the experience of the majority of
commercial growers indicates that this is unnecessary. When
orchards are well cultivated, so as to give a minimum of protection to the puparia through the winter, and when a good system of spraying is practiced, the apple maggot is not troublesome.
It is probable that under these conditions most of the flies are
killed by the arsenate of lead used for the control of the codlin-
moth.

REFERENCES

Maine Agr. Exp. Sta., Ann. Rept., 1889, pp. 190–241.
Maine Agr. Exp. Sta. Bull. 109.   1904.
R. I. Agr. Exp. Sta., Ann. Rept., 1904, pp. 191–201.
U. S. Bur. Ent. Circ. 101, 1908.
Cornell Agr. Exp. Sta. Bull. 324.   1912.

THE APPLE CURCULIO

Anthonomus quadrigibbus Say

Apples are subject to attack by three species of weevils
which are, in the order of their importance, the plum curculio,
the apple curculio and the apple weevil. The first will be
discussed in detail under plum insects (page 243).

The apple curculio is generally distributed over the Eastern
states and Canada, where it breeds abundantly in wild crab
and thorn apples. As an apple pest its work has often been
confused with that of the plum curculio, for the two species
usually work together and deform the fruit in a similar manner.

It has come into prominence principally in Missouri and
southern Illinois, and as a rule only in connection with destructive outbreaks of the plum curculio. Conditions favoring the
presence of the one seem also to favor the abundance of the

other. Usually the apple curculios are greatly in the minority, and their presence serves merely to supplement the injuries inflicted by the other species. In Connecticut this species has been recorded as seriously injuring young peach trees by puncturing the twigs.

The apple curculio is a reddish-brown snout beetle, and may be distinguished from its relatives by having four distinct humps on the posterior declivity of the wing covers, two on each side (Fig. 39). The thorax is usually striped with three ash-gray lines and the front part of the wing covers are more or less grayish. The female is about ¼ inch in length, the male a little

smaller. The beak is over one half the length of the body in the female, is slightly curved downward and carried projecting obliquely forward and does not hang down like an elephant's trunk as in the plum curculio.

FIG. 39. — The apple curculio (× 4).

The beetles hibernate in grass, under rubbish and in other sheltered places, and appear on the trees soon after the petals fall. They begin to feed on the young apples as soon as they are as large as small peas, and the female begins egg-laying soon after. In feeding the beetle punctures the skin of the apple by means of the small jaws at the tip of the beak and then eats out a cylindrical cavity in the pulp as deep as the length of the beak. Growth is stopped around the puncture, the surrounding tissue hardens and a knotty deformity results.

The cavity excavated by the female for the reception of the egg is similar to the one made in feeding but is considerably enlarged at the bottom. After placing the oval white or yellowish egg, $\frac{1}{25}$ inch in length, at the bottom of the cavity the female seals up the small external opening with a drop of excrement.

In feeding and in excavating the egg-cavity very little of the skin is swallowed; it is merely torn back out of the way. This is one reason why it is so difficult to kill the beetles with an arsenical spray.

In southern Illinois oviposition extends from late May to about the middle of July, and the average number of eggs laid by each female is about 65. The eggs hatch in four or five days and the larva feeds on the pulp, becoming full-grown in about 20 days. When full-grown it is nearly one half inch in length, footless, and owing to the enlargement of certain segments on the back is so strongly curved that it is unable to straighten out. It pupates within the cavity in which it has fed and in about one week the beetle emerges. The new brood of beetles, unlike the plum curculio, feed very little, but go into hibernation by the first of August.

Remedial measures.

Spraying with an arsenical as is practiced for the codlin-moth will destroy a small percentage of the apple curculios, but extensive experiments in Illinois have shown that additional applications for the curculios do not kill enough to pay for the expense incurred. Curculios thrive in overgrown, crowded, unpruned and uncultivated orchards, and may best be controlled by remedying these conditions. The trees should be pruned so as to admit as much sunlight as practicable, the ground should be kept free from weeds and the trees should not be so close together as to shade the entire ground. A large proportion of the infested apples drop and the insect completes its development in the fallen fruit. It is necessary, however, that the fruit remain in the shade, since even a few hours of direct sunlight is fatal to both larvæ and pupæ. Where the ground has been kept clean and smooth it will pay to rake the windfalls out into the sun, where they will dry up. This should be done early in the season, for great numbers of curculios develop in small apples not larger than a pea. Wild thorn apples in hedges and wood lots adjoining orchards serve as centers of infestation, and should be destroyed. Grassy borders and driveways, and particularly stone fences and stone piles serve as excellent hibernating quarters for the beetles. Clean orcharding is the most effective preventive of curculio attacks.

REFERENCES

Riley, 3d Mo. Rept. pp. 29–35. 1871.
Ill. Agr. Exp. Sta. Bull. 98. 1905.
W. Va. Agr. Exp. Sta. Bull. 126. 1910.

The Apple Weevil

Pseudanthonomus cratœgi Walsh

While generally distributed over the eastern United States, this weevil has been reported as injuring apples only in West

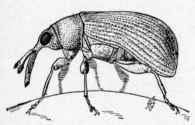

Virginia. Walsh in 1866 reared the beetle from a Cecidomyiid gall on the wild thorn.

The beetle is much smaller than the species last treated, being only $\frac{1}{10}$ inch in length. It is of a uniform light brown color and has the wing covers deeply striated but without

Fig. 40. — The apple weevil (× 15).

humps or tubercles of any kind (Fig. 40). The beetles emerge from hibernation in early spring. They feed on the foliage more or less throughout the season, but their principal food is the pulp of the fruit, which they obtain through minute punctures made in the skin.

The minute, yellowish-white oval eggs are deposited in cavities in the pulp eaten out by the females. The opening of the cavity is then sealed with a drop of excrement. The eggs hatch in 4 or 5 days, and the yellowish-white footless grub eats out a winding burrow in the fruit or may form a large irregular feeding chamber. The grubs are unable to develop in apples that continue to grow, being killed in many cases by the pressure of the proliferating plant cells. Under favorable conditions they become full-grown, and pupate in about 30 days on the average. The eggs are often laid in decaying fruit, and as many as 20 beetles have been reared from a single apple, although 4 or 5 is the more usual number. The pupæ occupy cells inside the fruit, and in a little over a week transform to beetles. The beetles remain about the trees during the remainder of the season, going into hibernation at the approach of cold weather. There is only one generation a year.

Remedial treatment.

The fact that the beetles feed more or less on the foliage makes it possible to kill them by spraying with arsenate of lead applied as for the codlin-moth. It has been shown in West Virginia that when the trees are treated in this way the injurious work of this weevil is completely prevented.

REFERENCE

W. Va. Agr. Exp. Sta. Bull. 126. 1910.

THE GREEN FRUIT-WORMS

Xylina antennata Walker
Xylina laticinerea Grote
Xylina grotei Riley

While very widely distributed throughout the United States and Canada, these three very closely related species have only occasionally attracted attention by their attacks on apples and other fruits. They caused considerable loss in Illinois and Missouri in 1870 and in New York in 1877, 1896 and 1913.

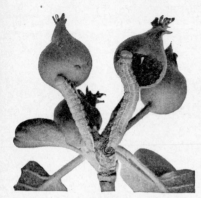

Fig. 41. — Green fruit-worms feeding on young apples.

They are more frequently found feeding on foliage of various forest trees, notably poplar, soft maple, hickory, wild cherry and boxelder. In addition to the apple they also attack the fruit of the pear, peach, plum, apricot, quince and currant. The green fruit-worms are large, light yellowish or apple green caterpillars (Fig. 41), with a narrow cream-colored stripe down the middle of the back, a wide cream-colored stripe along each side and many similarly colored mottlings or spots which sometimes form quite distinct stripes along the body above the broad lateral stripes. When fully grown they range from one to one and a half inches in length. They work during May and the first half of June. When young they feed upon the foliage or buds so that when the fruit is large enough for them to eat they are found to be about half grown. The caterpillars do not bore into the fruit, but usually begin eating on one side and often continue feeding until nearly half of the fruit is eaten. They go from fruit to fruit, one caterpillar thus ruining several fruits; in some orchards they have been known to destroy over a quarter of the crop. If the cavity eaten in the apple is not too large, it may heal over, leaving a light brown corky scar.

Fig. 42. — Pupa of the green fruit-worm.

The green fruit-worms do most of their damage to the young fruits in May, but some of them continue working until nearly the middle of June. During the first week in June most of the caterpillars get their full growth and burrow into the soil beneath the trees to a depth of from one to three inches. Here they roll and twist their bodies about until a smooth earthen cell is formed. Most of them then spin about themselves a very thin silken cocoon; some spin no cocoon. Soon after building the cocoon or earthen cell the caterpillar transforms to a dark

Fig. 43. — Green fruit-worm moth, *X. antennata* (× 1¾).

brown pupa (Fig. 42). In about three months, or about the middle of September, the moths (Fig. 43) emerge and go into hibernation in sheltered nooks; some of the pupæ, however, do not transform till early the following spring. The moths appear on the trees in March or April, and deposit their nearly globular, distinctly ridged, yellowish eggs singly on the bark of the smaller branches (Fig. 44).

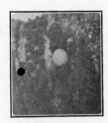

Fig. 44. — Green fruit-worm egg on apple twig, greatly enlarged.

Means of control.

As the green fruit-worms are about half grown when they begin feeding on the fruit it is then a very difficult matter to kill them with a poison spray. Earlier, when feeding on the buds and newly opened leaves, many of the young worms could doubtless be destroyed by a thorough application of arsenate of lead, 5 or 6 pounds in 100 gallons of water or dilute lime-sulfur solution.

Reference

Cornell Agr. Exp. Sta. Bull. 123. 1896.

CHAPTER III

APPLE INSECTS — BUDS AND FOLIAGE

The Bud-moth

Tmetocera ocellana Schiffermüller

This is one of the most numerous, destructive, and wide-spread of the insects attacking the opening buds of fruit trees. It is a European insect which has been ravaging American orchards for nearly a century, and is now more or less injurious yearly from Nova Scotia through Canada and the northern half of the United States to Oregon and Washington. Infested nursery stock is the principal source of new infestations. Working in the opening buds, it often " nips in the bud " a prospective crop of fruit, and it is especially destructive on recently budded or grafted trees and nursery stock. It more commonly infests apple trees, but pear, plum, cherry, quince, and peach trees and blackberry bushes are also attacked and sometimes seriously injured.

The half-grown, dark brown, black-headed caterpillars hibernate in obscure little silken hibernacula on the bark of the twigs, usually near the buds. Early in the spring, or as soon as the buds begin to open in April or May, these caterpillars leave their hibernating quarters and get into the opening buds (Fig. 45), where they feed upon the central expanding leaves and flowers, tying them together with silken threads. The petiole of one of the leaves is often nearly severed, and the edge of the wilting leaf is then rolled into a tube lined sparsely with silk in which the caterpillar lives for 6 or 7 weeks, going out to feed mostly at night. It often draws other leaves toward it and fastens them, thus forming a sort of nest. Some of the partly eaten leaves soon turn brown, thus rendering the work of the insect quite conspicuous. Where terminal buds are attacked the caterpillar sometimes burrows down the shoot for 2 or 3 inches, causing it to die.

The mature, nearly naked caterpillar is about half an inch in length, and of a cinnamon-brown color, with the head, thoracic shield and true legs black. Becoming full-grown in June, the caterpillars transform, and ten days are spent as brown pupæ in silken lined cocoons formed of leaves either rolled or tied together in the nests. The moths emerge over a period of six weeks, from June 5 to July 15 in New York. The dark ash-gray moths, with a broad,

Fig. 45. — Opening apple bud infested with a bud-moth caterpillar, showing the brownish particles thrown out at the tip by the larva.

cream-white band across the front wings, which have an expanse of $\frac{3}{5}$ of an inch, are night-flyers and closely mimic the

Fig. 46. — Bud-moth. Knight photo (× 4).

bark when at rest (Fig. 46). A few days after emerging, the females lay minute, flattened, disk-like, oval, nearly transparent, smooth eggs either singly or in small overlapping clusters on the leaves (Fig. 47). In a week or ten days a little black-headed, greenish caterpillar hatches, makes a silken tube open at both ends and sallies forth to feed upon the skin and inner tissues of the leaf, usually on the underside along the midrib. A thin protecting layer of silk is spun

over their feeding grounds, and the skeletonized portion of the leaf soon turns brown. Oftentimes when working on the underside of a leaf that touches an apple the little caterpillar eats into the fruit in several places, causing a blemish in the mature apple as shown in Figure 48. Most of this injury is usually attributed to the summer brood of codlin-moth larvæ. Turning brown in color in a day or two, the little caterpillars continue to feed on the leaves during July, August and a part of September, molting 3 or 4 times and getting about half grown. Some of them leave the foliage and go into winter

FIG. 47. — Eggs of the bud-moth. Greatly enlarged.

quarters on the twigs early in August. Before the leaves drop all are snugly tucked away in their very obscure silken winter homes about $\frac{1}{8}$ of an inch in length and covered with bits of dirt or sometimes made under a convenient piece of dead bud scale. There is thus but a single brood of the bud-moth annually.

Five little parasites work upon this insect in Europe, and at least three parasites (*Phytodietus vulgaris*, *Pimpla* sp. and *Microdus laticinctus*) help considerably to

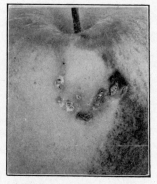

FIG. 48. — Apple injured by young bud-moth caterpillars in August.

check it in America. Birds also get some of the brown caterpillars, and a large muddauber wasp, *Odynerus catskillensis*, sometimes stores its cells with them to serve as delicious morsels for its baby grubs when they hatch.

Remedial treatments.

This bud-moth is a difficult insect to control. In nurseries or young orchards it is often practicable to go over the trees in May, when the " nests " are rendered quite conspicuous by one or two brown, dead leaves, and either pick off and destroy the nests, or crush them on the trees with the fingers so as to kill the inclosed caterpillars or pupæ. Skillful and thorough work with a poison spray will also control the bud-moth. Make two applications of arsenate of lead, 4 pounds in 100 gallons of water ; the first when the flower clusters first appear, and the second just before the blossoms open. Many growers add the poison to the lime-sulfur used against the scale and blister-mite, making the application just as the tips of the bud: begin to show green, and thus avoid a separate spraying for the bud-moth. Recent experiments, however, have shown that this early application of the poison has little effect. If these applications are thoroughly made and followed by the spraying usually given for the codlin-moth just after the blossoms drop, this pest can be effectually controlled.

REFERENCES

Cornell Agr. Exp. Sta. Bull. 50. 1893.
Cornell Agr. Exp. Sta. Bull. 107, pp. 57–66. 1896.

THE FRINGED-WING APPLE BUD-MOTH

Holcocera maligemmella Murtfeldt

Since about 1895 the light, greenish-yellow caterpillars, about $\frac{1}{5}$ of an inch long with a black head and thoracic shield, of this satiny, brownish-buff Tineid moth have been more or less injurious in apple orchards in western Missouri and adjoining states. The moths, which measure across expanded wings about $\frac{5}{8}$ of an inch, emerge from the ground early in April, lay their light yellow, oval, roughened eggs singly in the opening buds, and in a week or two the caterpillars appear. They fasten together some of the expanding leaves, feed within and work their way down into the center of the base of the open flower and leaf buds and developing shoots, often causing them to break off near the base. Thus the growth of the shoot is stopped, the terminal leaves killed and the prospective crop of fruit destroyed.

In about four weeks the caterpillars get their growth, crawl into the ground an inch or two, transform in a delicate, white, silken cocoon through tiny, brown pupæ to the moths which appear about the middle of July and lay eggs on the leaves for a second but less destructive brood of caterpillars which work in and kill the terminal leaves and buds on the new shoots. Becoming full-grown late in August, the caterpillars transform to pupæ in the ground and hibernate in that stage.

To control this fringed-wing apple bud-moth requires the most thorough work with a poison spray before the blossoms open, beginning just as soon as any green shows on the buds. Three applications of Paris green (1 pound in 100 gallons, with 3 pounds lime) have given good results. Two thorough sprayings with arsenate of lead before blossoming would doubtless prove equally effective.

REFERENCE

Mo. Agr. Exp. Sta. Bull. 42. 1898.

The Apple Bud-worm

Exartema malanum Fernald

This insect first attracted attention as an apple pest in northern Illinois in 1881, and has rarely been injurious since. The young caterpillars or bud-worms hatching from eggs laid singly on the terminal buds, are yellowish-white, tinged with pink or green. They devour the buds, then forming sort of a burrow by fastening a lower leaf-stalk to the branch; they feed upon this leaf and the newly formed wood, sometimes burrowing into it a short distance. In a short time this burrow is deserted, and the caterpillar, now of a dark flesh color, with polished black head and cervical shield, constructs a short, yellowish, woolly tube or case on the leaves, from which it sallies forth to another. When full-grown in June the caterpillars are about ½ an inch long, and they transform in their cases to the adult insect, a Tortricid moth. There is but one brood annually. The front wings of the moth expand about ½ an inch, and at the base and tip of each there is a large grayish-brown spot, those at the base being darker and the others mottled with white; between these spots the wings are white, with silvery reflections.

The tips of infested branches die back to the base of the first perfect leaf, and the trees present a blasted appearance. Apple trees only are attacked by this bud-worm, and it has a relentless enemy in the form of the small parasitic fly, *Microdus earinoides*.

As their work on the terminal buds and leaves is quite conspicuous, many of the young caterpillars can be destroyed in their burrows by pruning off and burning the infested tips. Later, in June, when they are feeding from their woolly cases on the leaves, a poison spray would prove an effective treatment.

THE CIGAR-CASE-BEARER

Coleophora fletcherella Fernald

About 1890 this interesting little case-bearer first appeared in injurious numbers in apple, pear and plum orchards in New York and Canada, where it continues to do more or less damage each year. It has been reported from New Mexico, Montana, Maine and British Columbia, where it was doubtless introduced on nursery stock. On account of its small size and peculiar habits, the insect itself will rarely be seen by the fruit-grower, but the curious little cigar-shaped suits in which the caterpillars live in May and June are quite conspicuous on the foliage.

The caterpillar protrudes itself from its case, eats a tiny round hole through the skin of the leaf and then mines out the interior tissue as far as it can reach and still retain its case. Thus many small, blotch mines are made which soon turn

Fig. 49.—Hibernating cases of the cigar-case-bearer. Herrick photo. Enlarged.

brown, and often whole leaves are thus mined out. About September 15, the insect goes into hibernation as a minute, half-grown caterpillar in a tiny curved case attached to the twigs (Fig. 49). About the middle of April, the caterpillars move and proceed to eat holes in the opening buds, the expanding leaves, the stems of the flowers and fruits and the young fruits (Fig. 50). Additions are built on to the winter case, but in about a month the dark, orange-colored, black-headed caterpillar, scarcely $\frac{1}{5}$ of an inch long, deftly makes, by mining and cutting out a cigar-shaped area of the leaf, its larger cigar-shaped case or suit. In the latter part of June the caterpillars cease feeding, securely fasten the cases to the leaves or branches, and in about three weeks transform within through light brown

Fig. 50. — Cigar-case-bearer attached to young pear which also shows several scars made by other case-bearers.

pupæ to the tiny steel-gray moths, with a wing expanse of about $\frac{3}{8}$ of an inch (Fig. 51). The females soon lay minute, yellow, pitted eggs among the hairs on the young leaves. Hatching in about two weeks, the tiny caterpillars work as miners in the leaves for two or three weeks, then construct their curious little curved cases from bits of the skins of the leaves, and by the middle of September migrate to the twigs, where they hibernate.

This cigar-case-bearer is capable of doing much damage to the young fruit and the foliage of fruit trees, and as it is protected by a case and mines in the leaves, it requires s k i l l f u l a n d thorough spraying to successfully control it. Possibly the strong sprays used against the San José scale in winter might reach the hibernating caterpillars in their tiny curved cases on the twigs. Early in the spring, or soon after the buds open and the caterpillars begin work, a thorough application of kerosene emulsion, diluted with 9 parts of water, has proved effective in Canada. In the commercial orchards of western New York case-bearers are usually controlled by the use of arsenate of lead, as recommended for the bud-moth, page 42.

Fig. 51. — Cigar-case-bearer moth (× 7).

REFERENCES

Cornell Agr. Exp. Sta. Bull. 93. 1895.
Fletcher, Rept. Ent. Ottawa, for 1894, pp. 201–206. 1895.
U. S. Bur. Ent. Bull. 80, Pt. II. 1909.

THE PISTOL CASE-BEARER

Coleophora malivorella Riley

This interesting insect spends about seven months of its life (from about September 1 to April 1) in hibernation as a minute, half-grown caterpillar in a small, pistol-shaped case

E

about ⅛ of an inch long attached to the twigs of its food-plants, which are the apple especially, but also include the quince,

plum and cherry. Early in April the little cases move and the caterpillars bore into and devour the swelling buds, expanding leaves, and especially the flowers. For four days about May 1, the cases may be found again fastened to the twigs while the caterpillars are molting inside. Unlike the cigar-case-bearer, this insect does not make a complete new case as it grows, but simply makes silken additions to the ends and sides of the old case. Most of their feeding is done openly and not

FIG. 52. — Pistol-case-bearers attached for pupation.

as miners, irregular holes being eaten in the leaves, often skeletonizing them. The caterpillars never leave their cases, but project themselves out far enough to get a foothold, then begin

to eat, holding the case at a considerable angle from the leaf. They are most destructive on the flowers, where they eat the petals and stems, thus destroying the prospective crop. Sometimes

FIG. 53. — Pistol-case-bearer moth (× 5).

they also bore into the young fruits. In the latter part of May they cease feeding, securely fasten the pistol-shaped cases, now about $\frac{1}{4}$ of an inch long, to the smaller branches (Fig. 52). These dark brown, bark-colored cases are made largely of silk,

particles of excrement and the pubescence of the leaves. Turning around in its case, the orange-colored caterpillar with black head, thoracic shield and legs, transforms in about a month through a light brown pupa to a pretty, little, dark drab-colored Tineid moth (Fig. 53) with a wing expanse of about $\frac{1}{2}$ an inch. It emerges from the "handle" end of the pistol (Fig. 54). On the basal third of the front wings white scales predominate in the fe-

Fig. 54. — Pistol-case-bearer, moth resting on its empty case.

males especially, the legs and antennæ have alternating rings of dark and white scales, and there is a conspicuous tuft of scales on the basal joint of each antennæ. The females glue

Fig. 55. — Eggs of the pistol-case-bearer. Greatly enlarged.

their minute, pretty, cinnamon-colored, inverted cup-like, ribbed eggs to the surface of the leaves (Fig. 55). The tiny caterpillars hatch from these eggs in about a week, or late in July, and eat little holes in the leaves. They soon construct of silk and excrement little pistol-shaped cases to which they build additions until about September 1, when they begin to migrate to the twigs and there fasten their cases to the bark and hibernate until April.

Since 1877 this insect has been very injurious in apple orchards at intervals of several years in New York and Pennsylvania. It is widely distributed in Canada and the United States, extending westward to New Mexico, where it was doubtless carried on nursery stock. It is capable of very destructive work and sometimes becomes so numerous that there is a casebearer for every two or three buds. Two or three tiny parasites help to control it.

Experiments show that this pistol-case-bearer can be effectually controlled with two applications of a poison spray before the blossoms open, as recommended for the bud-moth, page 45.

REFERENCES

Lintner, 1st Rept. State Ent., N. Y., pp. 163–167.   1882.
Cornell Agr. Exp. Sta. Bull. 124.   1897.
N. Y. (Geneva) Agr. Exp. Sta. Bull. 122.   1897.

THE PALMER-WORM

Ypsolophus ligulellus Hübner

This little Tineid moth occurs throughout the eastern half of the United States, but has done noticeable injury only in New York and the New England states. It is a remarkable example of the " ups and downs " of insect life, as it suddenly appears over a large area, does much damage for a year or two, then as suddenly disappears, often remaining in obscurity for half a century or more. Its first outbreak occurred in 1791 in New England, followed 62 years later by the second over a wider territory, including New York in 1853, and after waiting 57 years, or in 1900, it again ravaged apple orchards in New York. Its favorite food seems to be the foliage and fruits of apple, oak foliage, and, curiously enough, the spongy oak-apple galls are sometimes eaten. The caterpillars or palmer-worms are only about $\frac{1}{2}$ of an inch in length when full-grown, and of a general olivaceous or brownish-green color, usually with a light brown head; some are darker in color, with nearly black heads and thoracic shields. Two lateral and two dorsal whitish stripes give the dorsum the appearance of being traversed by two broad, dark stripes and a similar narrower mesal stripe of the general body color (Fig. 56). The active little palmer-worms move with a wriggling motion when touched, and work on the foliage and

Fig. 56. — The palmer-worm in its scanty silken net (× 5)

young fruits for 3 or 4 weeks in June. They skeletonize the foliage and spin down when disturbed, but are not loopers or measuring-worms, and although only about half as large as

Fig. 57. — Young apples eaten into by palmer-worms. Note the caterpillars in the first and third apples from the right.

canker-worms, may be more destructive, as they frequently eat large, deep holes into the sides of the young fruits (Fig. 57). They feed openly on the leaves, sometimes fastening 2 or 3 leaves together with silk, or, often rolling the edge of a leaf slightly, fasten it with silk threads and feed beneath this partial protec-

tion. Late in June, these palmer-worms transform to tiny brown pupæ (Fig. 58), which may be attached by a few silken threads, at their posterior ends, to the injured leaves, or may be found on the ground. In about ten days there emerges from these pupæ the minute gray or grayish-brown moths with a wing expanse of little more than $\frac{1}{2}$ an inch (Fig. 59). They are quite variable in color and markings. The front wings are more or less sprinkled with black scales and marked with 4 small, black spots arranged obliquely near the middle

FIG. 58. — Palmer-worm pupæ.

and 6 or 7 black dots near the fringed edge. The hind wings are heavily fringed and of a dusky color with glossy azure-blue reflection. There is but a single brood of palmer-worms annually; the moths which emerge early in July apparently hibernate and lay their tiny, delicate, pearly-white, oblong-oval eggs in the spring. The eggs are laid on the under side of the opening leaves in

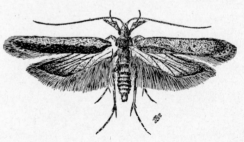

FIG. 59. — Palmer-worm moth; the wings on each side represent a different variety (× 5).

May; they are tucked away among the hairs or deposited in the angles of the veins.

Climatic conditions undoubtedly have much to do with the " ups and downs " of this insect. Excessively dry weather in April and May favors the development, and heavy rains often knock many from the leaves in June, and they never get back. Many of the palmer-worms are also killed by the maggot of a tiny parasitic fly known as *Apanteles* sp. Palmer-worms can be controlled with one or two applications of arsenate of lead, 4 pounds in 100 gallons of water or dilute lime-sulfur solution. The application should be made at the first appearance of the caterpillars in early June.

References

Cornell Agr. Exp. Sta. Bull. 187. 1901.

N. Y. (Geneva) Agr. Exp. Sta. Bull. 212, pp. 16–22. 1902.

Click-beetles

Corymbites caricinus Germar
Corymbites tarsalis Melsheimer
Corymbites cylindriformis Herbst

These three slender brown click-beetles, from $\frac{3}{8}$ to $\frac{1}{2}$ an inch in length, sometimes swarm on to apple and pear trees in the spring and attack the opening buds and the flowers, denuding orchards of blossoms in some instances. Thus far their ravages have been confined to Nova Scotia, British Columbia and other localities in Canada. In their larval stage as wireworms, these click-beetles doubtless bred on the roots of grasses, either in the orchards if in sod or in near-by fields, so that thorough cultivation in late fall would help much to reduce their numbers. As the beetles drop quickly and feign death when the tree is jarred, they can be readily captured on sheets or plum curculio-catchers. Or many of them would doubtless succumb to a strong poison spray of arsenate of lead, 6 or 8 pounds in 100 gallons.

Reference

Fletcher, Rept. Exp. Farms Ottawa 1895, p. 149, 1896.

The Ribbed Cocoon-maker of the Apple

Bucculatrix pomifoliella Clemens

The small but conspicuous whitish, distinctly ribbed cocoons, about ¼ of an inch long, of this insect, often occur in large numbers in autumn on the undersides of the smaller branches

of apple trees (Fig. 60), and may first reveal to the fruit grower its presence in the orchard. It is usually a local pest in widely separated

Fig. 60. — Cocoons of the ribbed cocoon-maker.

orchards, but has a wide distribution over the eastern half of Canada and in the United States from Maine to Texas. Although it is capable of doing much damage to foliage, it is rarely a serious pest, and attacks only the apple, other fruit trees being apparently immune even when growing in close proximity to apples. A little brown pupa hibernates in the white cocoons, and when the leaves are unfolding in May it works halfway out of one end of the cocoon and there emerges a tiny, light brown moth about ¹⁄₁₀ of an inch in

Fig. 61. — Ribbed cocoon-maker moth (× 8).

length with each front wing marked with a large, dark brown spot (Fig. 61). In a few days, minute, pale green, elliptical, iridescent, roughened eggs are laid singly on the under surface of the leaves. The tiny caterpillars which hatch from these eggs in from 6 to 10 days burrow directly into the leaf, where for about a week they make narrow mines nearly $\frac{3}{4}$ of an inch in length. It then comes out of this mine and makes a thin, white, silken molting cocoon (Fig. 62), within which its skin is shed in a few days, and the greenish-brown caterpillar feeds openly on the surface of the leaves near the edges. A second larger molting cocoon is made in about 4 days. Two days later the caterpillars appear in their last stage and proceed to skeletonize the foliage for about a

Fig. 62. — Apple leaf showing mines and "molting cocoons" of the caterpillars, enlarged.

week, finally wandering about to find a suitable place to spin their true ribbed cocoons early in July on the leaves, young fruits or twigs and larger limbs. It requires 3 or 4 hours to build one of these ribbed cocoons (Fig. 63). The pupal stage lasts from 1 to 2 weeks in these summer cocoons, most of the moths emerging by August 1 in New York. The mines of the young caterpillars are usually near the centers of the leaves, while most of the skeletonizing is done near the edges and always on the upper surface. When badly eaten, the leaves turn brown and curl. The caterpillars often hang suspended from the leaves by silken threads, and may be blown on to other food-plants near by, where their cocoons are sometimes found.

In Maine there is apparently but one brood of this insect annually, as the moths did not appear until spring from pupæ formed August 1. In southern New York, however, there are two distinct broods, the moths emerging in July lay eggs early in August and the second brood of caterpillars work on the leaves during August and September in the same manner as the first brood. Being more numerous, the work of the second brood is

Fig. 63. — The ribbed cocoon-maker building its cocoon. Enlarged.

usually more conspicuous and extensive than that of the spring brood. Eighty per cent of the cocoons made in autumn are within two feet of the ends of the branches on the lower third of the trees.

The summer brood is far less liable to be attacked by parasites and diseases, moths emerging from 80 to 90 per cent of the cocoons in July, whereas it is unusual for moths to emerge from more than 50 per cent of the winter cocoons. Five tiny hymenopterous parasites, *Cirrospilus flavicinctus*, *Encyrtus bucculatricis*, *Mesochorus politus*, *Apanteles cacœciœ* and *Zaporus* sp., play an important part in checking this pest. Many of the cocoons are often stripped from the twigs by birds in winter, and in summer the caterpillars are frequently found in webs spun across the surfaces of the leaves by the spiders, *Dictyna foliacea* and *Araneus displicatus*. There is also a great mortality, sometimes nearly 50 per cent, among the larvæ and pupæ in the cocoons in autumn, due apparently to some disease, causing them to shrivel and become dry and hard.

Remedial measures.

Sprays of whale-oil soap (1 pound in 1 gallon of water), the lime-sulfur wash, and probably the miscible oils (1 gallon in 10 of water) applied thoroughly at any time during the winter or early spring, while the tree is dormant, will soak through the white cocoons and kill the hibernating pupæ. Judicious pruning and burning of infested twigs in winter would destroy many. The application of a poison spray in the latter part of June will kill many of the tiny caterpillars then feeding on the leaves.

REFERENCE

Cornell Agr. Exp. Sta. Bull. 214. 1903.

THE LESSER APPLE LEAF-ROLLER

Alceris minuta Robinson

Throughout the eastern United States a small, pale yellowish-green caterpillar, about $\frac{1}{2}$ of an inch long, with a yellow head and thoracic shield, often draws the opposite edges of apple leaves together upwards and fastens them with silk. Living within the shelter of this folded leaf it feeds over the inner surface, often partially skeletonizing the leaf and causing it to turn brown. Sometimes nursery stock and young orchards are so badly infested, a majority of the leaves being folded and brown, that from a distance the trees appear as if a fire had swept through them, and much injury results. Older bearing trees are rarely seriously damaged by the insect. Besides being a serious menace to young apple trees, and sometimes attacking pears also, it is the yellow-headed cranberry worm, a destructive pest of cranberry bogs (see page 462).

There are three generations or broods of this insect annually on apple in the latitude of Missouri, but it is two-brooded in the cranberry bogs of Massachusetts. The moths are dimorphic; those of the one or two summer broods have bright, orange-colored front wings, while those of the autumn brood have slaty-gray front wings, often with a dusting of orange scales. The moths have a wing expanse of about $\frac{3}{4}$ of an inch. The third or dimorphic brood, bearing the varietal name of *cinderella*, appear in September and October, and hibernate in sheltered places. In the spring these gray moths emerge from their winter quarters and lay their minute, disk-like yellow eggs on the unfolding young leaves. Two or three broods of the caterpillars fold the leaves during the growing season, the first working in May, the second in July, and the third brood, where it occurs, works in August or September.

A single folded apple leaf usually furnishes sufficient food for a caterpillar. After feeding for 3 or 4 weeks, it spins a delicate, silken web or cocoon within the leaf, and then transforms to a small brown pupa about $\frac{1}{4}$ of an inch long, and characterized by a curious knob-like projection from the front of the head. In a week or ten days these pupæ work their way about half out of the folded leaves, and the moths emerge.

This apple leaf-folder has many parasitic enemies which are often effective aids in restricting its injuries. A Tachina-fly and several hymenopterous parasites are known to prey upon the caterpillars, and in Michigan a flock of birds, probably the rusty grackle, have been seen destroying many of the pupæ, neatly picking them out of the folded leaves.

Remedial measures.

It is often practicable in nurseries and young orchards to employ men or boys to go through and pinch the folded leaves, thus destroying the caterpillars within. This should be done

early in the season when the first brood is working, so as to prevent the development of later broods.   The insect can also be
controlled with a poison spray.   Spray thoroughly just as the
eggs are hatching early in the spring, when the first leaves are
unfolding, with arsenate of lead, 4 to 6 pounds in 100 gallons of
water.   Repeat the application in a few days if necessary.   If
the first brood is missed or not treated, quite effective work can
also be done against the later broods of caterpillars with the
poison sprays if applied just as the eggs are hatching.

REFERENCES

Forbes, 4th Rept. Ent. Ill., pp. 75–85.   1889.
Mo. Agr. Exp. Sta. Bull. 36, pp. 63–73.   1896.
Iowa Agr. Exp. Sta. Bull. 102.   1909.

THE APPLE LEAF-SEWER

Ancylis nubeculana Clemens

This insect works on apple leaves in much the same way as
the lesser apple leaf-folder, but it differs much in its life history.
The greenish-yellow caterpillar is about $\frac{1}{2}$ an inch long when full-
grown and has a yellowish head and cervical shield, the latter
with a conspicuous black spot near each outer hind corner.
These caterpillars hatch in early June, soon fold over a portion
of a leaf, and finally draw the opposite edges of the whole leaf
together upwards and securely fasten or sew them with silk.
Within this hollow shelter they feed on the green tissues during
the rest of the season, becoming full-grown in autumn.   Their
leafy home is then lined with silk and the caterpillars hibernate
therein in the fallen leaves.   There is thus but a single brood
annually, the caterpillars transforming through yellowish-brown
pupæ to the moths in about 10 days in April.   The pupæ wriggle
their way through the back of the decayed leaves and the moths
continue to emerge for about a month.   The oddly marked front

wings of these Tortricid moths are white with brown mottlings and shades, and have an expanse of about ¾ of an inch.

This leaf-sewer rarely does serious injury, but in a few instances it has appeared in alarming numbers in orchards in western New York and in Ontario, Canada.

Well cultivated orchards rarely suffer from this pest, as most of the hibernating caterpillars are buried with the fallen leaves. The effective method of raking up and burning the leaves in autumn would be practicable in some cases. A thorough application of a poison spray in early June will also kill many of the caterpillars then just beginning work on the foliage.

The Fruit-tree Leaf-roller

Archips argyrospila Walker

Recorded as common throughout practically the whole of the United States, this insect is one of the most destructive of the

FIG. 64. — Egg-mass of the fruit-tree leaf-roller on an apple twig, enlarged. Herrick photo.

leaf-rollers infesting fruit trees. It has been especially injurious in the orchards of New York, Missouri and Colorado, attacking apple, pear, cherry, plum, apricot, quince, rose, currant, raspberry and gooseberry, besides about a dozen different kinds of forest trees; curiously enough peaches seem to be exempt from attack. It is thus a very general feeder, and it sometimes strips fruit trees and ruins many of the young fruits.

The eggs are laid in June on the bark of the twigs in small flat, light brown or grayish patches, each patch containing about 150 eggs and covered with an impervious gummy substance (Fig. 64). The winter is passed in the egg stage. The caterpillars hatch about May 1, and enter the opening buds, where they roll and fasten the leaves loosely together with silken threads into a nest within which they feed (Figs. 65 and 66). After the fruits set, they are often included in the nests and ruined by the caterpillars eating large irregular holes in them (Fig. 67). The caterpillars get their growth in 2 or 3 weeks, and are then about ¾ of an inch long, light green in color with the head,

FIG. 65. — Larva of the fruit-tree leaf-roller beginning its nest on an apple leaf. Knight photo.

legs and thoracic shield varying from brown to black. About ten days are spent as a brown pupa (Fig. 68) in a delicate silken web or cocoon in the nest early in June. The moths emerge, and

FIG. 66. — Apple leaf rolled by the fruit-tree leaf-roller. Knight photo.

soon lay the peculiar patches of eggs on the bark, thus completing the life-cycle of the single annual generation of this leaf-roller. The pretty little moths, measuring about ¾ of an inch across the expanded wings, vary considerably in coloring and markings (Fig. 69). The front wings are rust-brown in color, marked with bands and spots of very pale yellow.

A number of hymenopterous parasites attack this leaf-roller. Toads often eat many of the caterpillars that drop from the trees, and red-winged blackbirds are efficient enemies in Colorado.

Means of control.

The leaf-roller has been found a difficult insect to control by ordinary spraying with arsenical poisons. Recent work in Colorado has shown, however, that over 95 per cent of the eggs can be destroyed by one thorough application

Fig. 68. — Pupa of the fruit-tree leaf-roller. Knight photo.

Fig. 67. — Small apples eaten by the leaf-roller. Knight photo.

of a miscible oil, one part in 19 parts of water, made early in the spring while the trees are dormant. Efficient work against the eggs can also be done with a 10 per cent kerosene emulsion

Fig. 69. — Fruit-tree leaf-roller moth. Knight photo ($\times 2\frac{1}{2}$).

whenever for any reason it is undesirable to use a miscible oil. In case the eggs have not been treated it is necessary to resort to arsenical sprays to kill the young caterpillars. Use arsenate of lead, 6 pounds in 100 gallons of water, making the first application just as soon

as the buds begin to burst, and the second when the blossom
buds in the cluster begin to separate. This should be followed
by the regular spray for the codlin-moth just as the last of the
petals are falling.

REFERENCES

Col. Agr. Exp. Sta. Bull. 19, pp. 3–9. 1892.
Mo. Agr. Exp. Sta. Bull. 71. 1906.
Cornell Agr. Exp. Sta. Bull. 311. 1912.
Col. St. Ent. Circular 5. 1912.
U. S. Bur. Ent. Bull. 116, Pt. V. 1913.

THE OBLIQUE-BANDED LEAF-ROLLER

Archips rosaceana Harris

The caterpillar of this leaf-roller varies from a light yellowish-
brown to apple green in color and is about ¾ of an inch in length,
with a brownish-black head and thoracic shield, the latter usually
with a whitish cephalic border, and often more or less green
bordered with black. It lives in a nest similar to that of the
bud-moth, formed by rolling or folding and tying together leaves
of apple, pear, cherry, plum, peach, rose, raspberry, gooseberry,
currant, strawberry, and several other trees, wild berries, red
clover, more than half a dozen weeds, and eats into growing
cotton bolls. The insect is common and widely distributed
throughout the northern United States, where it is sometimes
quite injurious, especially on apple trees. In addition to eating
the leaves and often checking the growth of shoots, the caterpillars sometimes attack the young fruits, either gnawing off the
skin or eating holes in them.

There are two broods of the caterpillars annually, the first
brood working in May and June and the second in July and
August.

When full-grown, the caterpillars transform in their nest to
brown pupæ, from which emerge in a week or ten days the

F

oblique-banded, light cinnamon-brown colored moths with a wing expanse of about an inch (Fig. 70). The front wings are reticulated with brown and each is crossed by three broad, oblique dark brown bands. The females lay their eggs in flat patches (Fig. 71) on the bark, and the insect hibernates in this stage.

Fig. 70. — The oblique-banded leaf-roller moth, from life. Knight photo ($\times 3\frac{1}{2}$).

Fig. 71. — Egg-mass of the oblique-banded leaf-roller. Herrick photo.

A large ichneumon parasite, *Glypta simplicipes*, destroys many of the caterpillars of this oblique-banded leaf-roller.

The Four-banded Leaf-roller

Eulia quadrifasciana Fernald

This smaller, bright yellow caterpillar, about $\frac{1}{2}$ an inch long, sometimes works on apple trees with and in much the same way as the oblique-banded leaf-roller. It webs a few leaves together and riddles them. The insect is widely distributed over the northern United States, and has done considerable injury in apple orchards in Delaware, working with the preceding species, and has helped to nearly strip trees in Canada. The moths are lemon-yellow, with the front wings reticulated with orange-red and each crossed by two narrow oblique darker bands. They measure across expanded wings about $\frac{5}{8}$ of an inch, and have been reared late in May and June.

These two Tortricid leaf-rollers can be controlled by the measures recommended for the fruit-tree leaf-roller on page 64.

The Apple Leaf-skeletonizer

Canarsia hammondi Riley

The small greenish or brownish caterpillars of this insect are about ½ an inch long, with 4 black, shining tubercles on the back just behind the head, and usually with a broad darker stripe along each side of the back. They live upon the upper side of the leaves under a thin web of silken threads, where they eat out the green pulpy portion, leaving a network of veinlets and giving the foliage a skeletonized and rusty or highly blighted appearance. There are two broods of the insect annually, one working in midsummer and the other, usually the more numerous, in September and October. Sometimes the cater-

FIG. 72. — The apple leaf-skeletonizer moth (× 5).

pillars work gregariously in a nest of several leaves webbed together. Pupation occurs on the leaves and the second brood hibernates in the pupa stage. This Pyralid moth (Fig. 72) has an expanse of about one half an inch; the front wings are glossy purplish-brown, each marked with two silvery-gray transverse bands.

This leaf-skeletonizer is most common in the Mississippi Valley, where it is sometimes quite destructive in nurseries and young orchards. It works practically only on apple trees, rarely attacking plum and quince.

Two or three parasites attack it. As the caterpillars feed openly on the surface of the foliage, they can be easily killed with a poison spray applied as early in the season as their skeletonizing work is noticed.

REFERENCE

Forbes, 4th Rept. State Ent. Ill., pp. 58–64. 1889.

THE LEAF-CRUMPLER

Mineola indigenella Zeller

Although widely distributed over the northern portion of the United States as far west as Colorado, and occurring also in Canada, this leaf-crumpler is usually most common and destructive in Missouri and the surrounding states. Nursery stock and young apple or quince orchards are sometimes seriously injured, and plum, cherry, peach and rarely pear may be included among its food-plants. The partially grown, reddish-brown caterpillars hibernate in a slender, blackish, oddly crooked or twisted horn-like or cornucopia-shaped tube or case nearly an inch long, hidden among several partially eaten brown and crumpled leaves, all fastened securely to the branches. Early in the spring the hungry caterpillars cut loose the fastenings of their winter home and travel with the odd-shaped cases to the opening buds and begin feeding. Several leaves are often fastened together, and sometimes the young fruits are attacked or the tender bark gnawed from the twigs.

When full-grown, in the latter part of May or early June, the caterpillars are about $\frac{3}{5}$ of an inch in length, and of a dark greenish-brown color, with the head and cervical shield dark reddish-brown; there is also a flattened blackish prominence on each side below the shield. After fastening the crooked cases to the bark and securely closing them the caterpillars transform in about 2 weeks through reddish-brown pupæ to Pyralid moths measuring about $\frac{7}{10}$ of an inch across the expanded wings (Fig. 73). The front wings are pale brown, with patches and streaks of silvery white. The eggs are said to hatch in about a week. The young, brown caterpillars feed

preferably upon the leaves of the tender shoots, and soon construct their peculiar cases of silk, excrement and other débris. Additions are built on around the larger end of the case as the caterpillar grows, and although rough exteriorly, it is smoothly lined with silk on the interior. As cold weather approaches, the cases containing the partially grown caterpillars are fastened up for hibernation. There is but a single brood annually.

At least three parasites attack this leaf-crumpler and render efficient aid in holding it in check.

Fig. 73. — The leaf-crumpler moth ($\times$ 3½).

As their hibernating quarters are rendered quite conspicuous by surrounding brown and crumpled leaves, it is practicable to hand pick these winter nests from the bare trees, and thus destroy many of the young caterpillars. Over 140 have been taken from a single 6-year-old tree in Oklahoma, and one man collected 1584 of the winter nests in 4 hours.

The insect can be easily controlled also by the thorough application of a poison spray just before and again just after the blossoming period of the trees.

REFERENCES

Forbes, 4th Rept. State Ent. Ill., pp. 65–74. 1889.
Mo. Agr. Exp. Sta. Bull. 36, pp. 73–80. 1896.

THE TRUMPET LEAF-MINER

Tischeria malifoliella Clemens

Over the eastern half of the United States and Canada this is often the most common and destructive of the insects which make mines in apple leaves. The full-grown caterpillars are flat-bodied, about $\frac{1}{5}$ of an inch in length, without legs, light green, with head, thoracic and anal shields brownish. Their brown, blister-like, trumpet-shaped mines about $\frac{1}{2}$ of an inch long are always made just beneath the epidermis on the upper side of the leaves, being scarcely noticeable from the underside unless the leaf is held to the light. The mines begin at the point where the minute, elliptical, greenish-yellow, disk-

Fig. 74. — Mines of the trumpet leaf-miner.

like, iridescent eggs are laid, the tiny caterpillars entering the leaf beneath the edge of the eggshell. The mines continue for a short distance as a narrow line, gradually growing wider, and then often suddenly expand into a broad, blotch mine, the whole having a trumpet-shaped appearance. The first half of the mine is usually crossed by crescent-shaped stripes of white (Fig. 74).

In the North there are two broods of this miner annually, the first brood of caterpillars working in June and July and

the second in August and September. Four broods have
been observed near Washington, the mines being made in May,
July, August and September; the life-cycle of a generation
was about 33 days. The transformation through the tiny,
greenish-brown pupæ occurs in the mines. In autumn the
mines are heavily lined with white silk and the little caterpillars pass the winter therein on the fallen leaves. The adult
insect is a tiny Tineid moth with shining dark brown front
wings, tinged with purplish and dusted with pale yellowish
scales (Fig. 75). They measure across expanded wings only
about ¼ of an inch.

Sometimes the
mines are so nu-
merous as to in-
volve much of the
leaf, which curls
and finally drops 2
or 3 weeks earlier
than usual, thus

Fig. 75. — The trumpet leaf-miner moth (× 10).

preventing the full development of the fruit and reducing
the vitality of the tree. Sixty-eight caterpillars have been
found working on a single leaf. On large trees, the foliage
on the higher branches is usually more seriously infested.

The favorite food-plant of this native leaf-miner is the cultivated apple, but it also breeds on native crab and haw trees.
The trumpet-shaped mines in the leaves of blackberry and
raspberry usually attributed to this insect are now considered
to be the work of two other species, the *Tischeria œnea* and
roseticola of Frey and Boll.

At least half a dozen tiny hymenopterous parasites destroy
many of these miners, thus doing much to prevent the insect
from increasing to destructive numbers.

Remedial treatment.

As this insect hibernates on the fallen leaves, many of them

can be destroyed by plowing infested orchards either in late
fall or early spring. Thoroughly cultivated orchards will
rarely suffer serious injury from this miner. Experiments
indicate that many of the larvæ and pupæ can be killed in the
mines by thorough applications of 10 or 15 per cent kerosene
lime emulsion. Just as effective work with less danger of injuring the foliage could doubtless be done with " Black Leaf
40 " tobacco extract, one pint in 100 gallons of water, adding 4
pounds of soap to each 100 gallons to make the liquid stick and
spread better.

References

Brunn, Cornell Univ. Exp. Sta., Second Rept., pp. 155–157. 1883.
Forbes, 4th Rept. State Ent. Ill., pp. 45–50. 1889.
Conn. (Storrs) Agr. Exp. Sta. Bull. 45. 1906.
U. S. Bur. Ent. Bull. 68, Pt. III. 1907.
Del. Agr. Exp. Sta. Bull. 87, pp. 3–9. 1910.

Some Lesser Leaf-miners of the Apple

The following four species of small caterpillars work as
miners in the leaves of apple, and while often numerous enough
to attract attention, they rarely appear in sufficient numbers
to do serious injury.

The spotted tentiform leaf-miner (*Lithocolletes blancardella*
Fabricius).

The tiny light yellow caterpillars, only $\frac{1}{5}$ of an inch long,
make a small mine
about $\frac{1}{2}$ an inch long
on the lower surface
of the leaf, which
causes a slight crimp-
ing of the leaf, thus
giving the mine a
tent-like appear-

Fig. 76. — The moth of the spotted tentiform leaf-
miner (× 12).

ance. From the upper surface, the mine has a spotted appearance due to the caterpillars not mining out the whole interior, but eating a little here and there in the mine. The mines are finished in September, the caterpillars transform to pupæ therein, and the winter is passed in this stage on the fallen leaves. The minute Tineid moths which emerge in the spring have golden brown front wings marked with white streaks and spots and a black apical spot (Fig. 76). This European miner is quite common on apple leaves in the eastern United States, but has not yet been recorded as doing serious injury.

REFERENCES

Brunn, Cornell Univ. Exp. Sta., Second Rept., pp. 148–150. 1883.

The unspotted tentiform leaf-miner (*Ornix geminatella* Packard).

The tentiform mines of this insect are larger, and distort the leaves more than those of the preceding species (Fig. 77). The grayish caterpillars, about $\frac{1}{4}$ of an inch long, have a row of 6 black spots across the head and 4 larger ones across the dorsum of the first thoracic segment. They eat the whole interior except the veinlets, so that the mine appears brownish but not spotted on the upper surface. When nearly full-grown the cater-

FIG. 77. — Mines of the unspotted tentiform leaf-miner.

pillars leave their mines, and rolling over the edge of the leaf
feed beneath for a short time, then line these retreats heavily
with a silken cocoon within which they pupate. There are
several broods each season. The tiny, dark, steel-gray moths
emerge in the spring and measure only $\frac{1}{3}$ of an inch across
the expanded wings. This miner is widely distributed
across the northern half of the United States, and is apparently more common than the spotted tentiform miner.
In some cases two-thirds of the leaves in orchards have been
distorted by from 2 to 4 of the unspotted mines, yet no very
serious injury resulted. The insect also attacks pear and wild
cherry foliage.

REFERENCES

Brunn, Cornell Univ. Exp. Sta., Second Rept., pp. 151–154.　1883.
Forbes, 4th Rept. State Ent. Ill., pp. 51–57.　1889.
N. Y. (Geneva) Agr. Exp. Sta. Bull. 180, pp. 131–134.　1900.

The serpentine leaf-miner (*Nepticula pomivorella* Packard).

The tiny, dark, emerald-green caterpillars, about $\frac{1}{10}$ of an
inch long, make narrow, tortuous or serpentine mines, often
2 inches in length and less than $\frac{1}{16}$ of an inch wide just beneath
the upper surface of the leaves of the apple and pear. The
first half or two thirds of the mine is broader and nearly filled
with a continuous zigzagging thread of black excrement.
The insect is quite common in Canada and the northeastern
United States. In October, the tiny green caterpillars are
sometimes seen hanging by silken threads from the leaves.
They soon find their way to the twigs, where they spin small,
oval, dense, brown cocoons about $\frac{1}{8}$ of an inch long on the back,
often in a crotch. These cocoons resemble, and could be easily
mistaken for, Lecanium scales. In May the caterpillars transform through brilliant green pupæ to the minute, shining, pur-
plish-black moths with tufted, reddish-yellow head, that emerge
early in June. Thus far no very serious injury has been recorded by this interesting little Tineid serpentine miner.

The resplendent shield-bearer (Coptodisca splendoriferella Clemens).

Throughout the northern United States, from Maine to Minnesota often there may be found attached to the bark of apple, pear, quince, thorn-apple and wild cherry trees curious little, oval, disk-shaped, seed-like, yellowish bodies about $\frac{1}{10}$ of an inch long (Fig. 78). From these little shields or cases, fastened to the bark at one end by a silken button, there emerges in May a tiny, brilliantly colored, golden-headed moth (Fig. 79). The basal half of the front wings are leaden-gray with a resplendent luster and the remainder

FIG. 78. — Hibernating cocoons of the resplendent shield-bearer.

golden with silvery and dark brownish streaks. These beautiful little creatures run about on the leaves in the sunshine and lay their eggs, from which hatch the tiny, light, yellowish-brown, legless caterpillars about $\frac{1}{8}$ of an inch in length. These make an irregular dark-colored blotch mine, about $\frac{1}{4}$ of an inch in diameter, in the leaves and observable from both surfaces. When full-grown, the caterpillars line a portion of the mine with silk, deftly cut it out and thus form their seed-like shield. Dropping from the leaves

FIG. 79. — The moth of the resplendent shield-bearer ($\times$ 10).

in July by a silken thread, they finally reach the bark or the ground, or are blown to other trees, where the cases are fastened. A second brood of the little miners works on the leaves in September and during October they fasten their cases to the bark and hibernate therein as caterpillars.

Several quite serious outbreaks of this tiny shield-bearer

have occurred at Washington, in Connecticut and in Michigan on apple, quince and wild cherry, sometimes 25 or 30 mines occurring in a single leaf. The bark of the trunk and larger branches were fairly covered with the hibernating cases, 47 having been counted on a spot not larger than a dime.

REFERENCE

Comstock, Rept. U. S. Ent. for 1879, pp. 210–213.

Natural enemies of the lesser leaf-miners of the apple.

All of these little leaf-miners have enemies which are more or less effective aids in preventing their occurrence in injurious numbers. At least two tiny Chalcid parasites, *Sympiesis nigrifemora* and *Astichus tischeriae* attack both the spotted and unspotted tentiform-miners. Ants often tear open the cases of the resplendent shield-bearer and devour the inclosed caterpillar or pupa; two minute hymenopterous parasites also attack this miner. We have bred a tiny parasite from the serpentine miner, and many of the scale-like hibernating cocoons have been found in the stomachs of chickadees.

Remedial treatments.

In well cultivated and thoroughly sprayed orchards, leaf-miners rarely become abundant enough to cause appreciable loss, and special treatment is therefore rarely necessary. As all of these lesser leaf-miners feed inside the leaves they cannot be effectively reached with poison sprays. For the tentiform-miners, resort to the treatments recommended for the trumpet leaf-miner. Possibly a spray of "Black Leaf 40" tobacco extract, 1 pint in 100 gallons of water, to which 4 or 5 pounds of soap have been added would penetrate mines and kill the caterpillars of the shield-bearer and serpentine-miner. When very numerous many of the cases of the shield-miner could be scraped from the trees and destroyed.

The Spring Canker-worm

Paleacrita vernata Peck

Canker-worms are among the oldest and most destructive of American orchard pests.   The term " cancker-worme " apparently originated in England in 1530, and was used for several different insects in the first authorized English version of the Bible in 1611.   As early as 1661, John Hull quaintly related that " the canker-worm hath for four years devoured most of the apples in Boston, that the apple trees look in June as if it was the 9th month" (meaning November).   Several other serious outbreaks of canker-worms were recorded in New England during the next century.   Although the scientific name of *vernata* was applied to these canker-worms in 1795, and another name, *pometaria*, was proposed for some of the moths in 1841, it was not demonstrated until 1873 that two quite different species of insects had been masquerading as *the canker-worm* in America for more than 200 years.

Canker-worms belong to the Geometrid group of moths whose caterpillars are called measuring-worms, span-worms or loopers from their peculiar manner of walking.   Although several kinds of these measuring-worms are destructive to fruits, often working in the same orchards, the term canker-worm is restricted to the two distinct species recognized in 1873 and then given the common names of the spring canker-worm and the fall canker-worm.

Both of these native species often occur together in injurious numbers in the same locality.   The apple and elm are favorite food-plants, although several other fruit and shade trees are often attacked.   The caterpillars appear on the trees in early spring and work mostly during May, skeletonizing the leaves, which soon turn brown.   In June badly infested trees or orchards often appear from a distance as though a fire had swept through them. Like many other insect pests, canker-worms have their "ups and downs," or periods of increase and decrease. Usually their destructive period lasts from 3 to 5 years or more before their enemies and climatic conditions succeed in bringing about the "down" period, which may last 10 or more years. The defoliated trees grow weaker each year,

FIG. 80. — Canker-worm moths caught on sticky band while ascending tree trunk.

mature but little fruit, and may finally succumb to the ravages of the hordes of canker-worms. If disturbed, the caterpillars often cling with their hind pro-legs and stand out straight, stiff and motionless, this strange attitude giving them a protective resemblance to the smaller twigs or leaf-stems; or they may drop suddenly and swing suspended in the air by a silken cord until the threatened danger has apparently passed, when they resume their normal looping positions or ascend their ropes somewhat sailor-fashion.

The male canker-worm moths have fully developed wings, while the females are practically wingless, only short stubs of wings being present. This lack of wings renders it necessary for the females to crawl up the trees to lay their eggs, and affords an opportunity to apply certain effective barrier remedial measures (Fig. 80). The moths are active only at night, and often ascend the trees in the greatest numbers between 7 P.M. and 10 P.M.

We have seen hundreds of the females ascending a single tree during the evening in a badly infested orchard; they are little disturbed by lights brought near. Many of them often secrete themselves in the crevices of the bark during the day.

Fig. 81. — Spring canker-worm, male moth (× 2).

There are striking and easily discernible structural differences between the spring and the fall canker-worms in all four stages of their life-cycles.

The spring canker-worm is common in Canada and in the northern United States from Maine westward to Kansas, thence southward through the Mississippi Valley to Texas. It also occurs in California. It is the species most often destructive in apple orchards, especially in its western range. The moths practically always emerge in early spring, sometimes during warm spells in February, but usually in March and April. They may continue to go up the trees for from 6 to 10 weeks. We have reared male moths varying in wing expanse from $\frac{5}{8}$ to

1¼ inches. The wings have a silky delicate appearance, the hind ones being unmarked and pale ash gray; on the slightly

darker, pale brownish-gray front wings usually three transverse, jagged, dark lines can be discerned, especially on the front edge of the wings (Fig. 81). The egg-laden wingless females are nearly ½ an inch long and of a rabbit-gray color, with white bands on the legs and usually a distinct black stripe along the middle of the back (Fig. 82). Both sexes of the moths have on top of each of the first seven abdominal segments two transverse rows of sharp reddish spines projecting backward; as there are no spines on the moths of the fall canker-worm it is easy to distinguish the two species of canker-worms in the moth stage.

FIG. 82. — Spring canker-worm, female moth (×2).

A single female moth of the spring canker-worm may lay over 400 eggs, which she tucks away in small, irregular pits or clusters in the crevices under bark scales and moss on the trunk and larger limbs of the trees. The delicate, slightly ridged, oval eggs are about $\frac{1}{50}$ of an inch long and of a dark iridescent metallic buff or purplish color (Fig. 83). The eggs hatch early in May, and the caterpillars feed for about a month on the pulpy portions of the leaves, thus skeletonizing them. The full-grown caterpillars are about an inch long, slender, and have but two pairs of pro-legs,

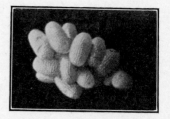

FIG. 83. — Eggs of the spring canker-worm. Enlarged.

thus being readily distinguished from the fall canker-worms, with their short third pair of pro-legs (Fig. 84). The spring

Fig. 84. — Full-grown spring canker-worms (× 1½).

canker-worms vary in color from a light mottled yellowish-brown to a dull black. Their more constant characteristics are a mottled head, a narrow yellow stripe just below the

G

spiracles, and a wide greenish-yellow stripe bordered by black lines along the middle of the venter (Fig. 85). Usually 3 narrow, more or less broken, yellow stripes can be distinguished ex-

tending along each side of the body above the spiracles.

By June 1, most of the caterpillars get their growth, spin down from the trees, and enter the ground an inch or more, where they

FIG. 85. — Spring canker-worm, side view (× 1½).

transform in a simple earthen cell to the greenish-brown pupæ (Fig. 86). There is but a single generation annually, the insect spending at least 9 months, including the winter, in the pupa stage.

Natural enemies of canker-worms.

Some of the caterpillars often fall a prey to several hymenopterous and dipterous (*Tachina*) parasites, predatory sucking bugs and ground-beetles. Potter-wasps sometimes store their clay nests with them, and other enemies are mentioned in the discussion of the fall canker-worm. But by far the most effective enemies of canker-worms are the birds. Over forty kinds of birds, especially the chickadees, thrushes and warblers, have been found feeding on the caterpillars, the eggs or

FIG. 86. — Pupæ of the spring canker-worm. Enlarged.

the egg-laden female moths. Yet in spite of the efficient aid thus rendered by the birds and other animal parasites, the cessation of destructive outbreaks of canker-worms is usually due more to certain obscure climatic or other local conditions. Early spring frosts often kill large numbers of the young caterpillars.

Remedial measures for canker-worms.

Although among the most destructive of the insect pests of orchards, canker-worms can be readily controlled. Well cultivated orchards are rarely injuriously infested with canker-worms. Thorough cultivation during June or later kills, or turns out for the birds or other enemies, many of the pupæ in their earthen cells or cocoons near the surface. Thus orchards can be kept practically free from the devastating hordes of these caterpillars by simply practicing one of the most essential factors in modern fruit-growing ; namely, thorough cultivation.

Barriers of various sorts have long been used to prevent their ascent or to trap and kill the wingless female moths when they attempt to crawl up the trunks of the trees to lay their eggs. To get the best results with these barriers it is necessary to know which species of canker-worm infests the trees. If it is the spring canker-worm, the barriers need not be applied until late in February or during March, depending upon the early occurrence of a warm spell. But in the case of the fall canker-worm the barriers must be put on late in October and kept in working order until the ground is well frozen ; in some of the more southern or warmer localities in this insect's range, where many of the moths often do not emerge until early spring, it will be necessary to maintain the barriers during March and April also.

Among the mechanical barriers, there are two simple devices that have been found effective. But few of the moths can get over a band of cotton batting several inches wide placed around the trunk, tied tightly with a string near the bottom edge, and the upper portion of the band then turned down over the lower, thus forming an inverted funnel-shaped barrier. When not kept matted down by frequent rains these cotton bands are very effective. A strip of mosquito wire netting, at least as fine as 16 wires to the inch, and about 14 inches wide, tacked so as to fit tightly around the tree at the top and held out from the bark for half an inch or more at the bottom by a spiral spring or nails driven into the tree, forms almost a perfect barrier to the wingless females. We have seen such wire traps nearly full of the moths in badly infested orchards. They should be crushed under the wire each night, and it is sometimes necessary to remove and empty the traps. While these mechanical barriers may be very effective in preventing the females from getting on to the tree to oviposit, they often lay many eggs below the barriers, and in the case of the wire traps, the young caterpillars may crawl through the meshes and reach the foliage; the cotton bands, if kept fresh and fluffy, would doubtless continue effective against the caterpillars. In applying such barriers all rough places on the bark must be smoothed or filled so as to allow no chance for the moths to crawl under the edge of the bands. Their effectiveness will depend largely on their proper application and maintenance in a good working condition while the moths are active.

Certain sticky band placed around the trunks of the trees are just as effective barriers as the wire traps or cotton bands, and they have the decided advantage of capturing and killing the egg-laden females, and also of effectively preventing the ascent of the young caterpillars that may be hatched below the bands. Among the sticky materials found effective are Tree Tanglefoot, a mixture of 5 pounds resin and 3 pints castor oil (some add Venice turpentine, 3 pints to this) and printer's ink mixed with black Virginia oil or some similar heavy oil to prevent its drying out too quickly. It is best to apply these on a band of tarred or other heavy paper 6 or 8 inches wide tacked or tied around the trunk and all roughnesses beneath filled with cotton. Keep the bands sticky by fresh applications when needed. On badly infested trees it is sometimes necessary to renew the bands or apply two, as many of the females may be able to cross the band over the dead bodies and wings of the males, which may completely cover the sticky portion. The use of these mechanical barriers or sticky bands are especially recommended on very large, rough-barked elms, or other trees that it would be difficult to spray or cultivate thoroughly.

If no effort is made to prevent the moths from ascending the trees and laying their eggs, either in the fall or spring, the voracious caterpillars can be killed with a poison spray in May. This has been demonstrated many times, but only the most thorough kind of spraying will conquer a hungry army of canker-worms. The most effective work can be done by applying the spray early, when the caterpillars are young and will thus succumb to a smaller dose of poison. Effective work has been done with Paris green at the rate of 1 pound in 100 gallons, but an arsenate of lead spray, 4 or 5 pounds in 100 gallons, has some advantages. For apple trees make one application just before the blossoms open and a second after the petals fall; if the work is thoroughly done, further applications will be rarely necessary. For shade trees, begin spraying as soon as the first leaves unfold or the young canker-worms are seen, which is usually early in May.

Under the modern system of fruit-growing, embodying thorough cultural and spraying methods, canker-worms will rarely find congenial conditions for their nefarious and destructive work; in fact they now seldom attract the attention of commercial growers.

REFERENCES

Peck, Nat. Hist. Canker-worm. 1796.

Harris, Insects Inj. Veg. pp. 332–343. 1841.

Ohio Dept. Agr. Bull. 2. 1903.

U. S. Bur. Ent. Bull. 68, Pt. II. 1907.

Conn. Agr. Exp. Sta. Rept. for 1907–1908, pp. 777–796. 1909.

The Fall Canker-worm

Alsophila pometaria Harris

Although not distinguished from the spring canker-worm until 1873, this fall canker-worm doubtless has been injurious in the New England states for half a century or more. It is now common in Canada and throughout the northeastern United States, extending westward into Ohio; and in 1891, orchards of apple, prune, plum, apricot and cherry were ravaged in western California, apparently by this eastern

Fig. 87. — Fall canker-worm, male moth (×2).

fall canker-worm. It is often destructive in apple orchards, sometimes working with the spring canker-worm or on neighboring orchard or shade trees; the two species have worked separately for several years on elm trees about a mile apart near Ithaca, N. Y.

The fall canker-worm is easily distinguished in all its stages from the spring species, and differs also in its life-habits. It derived its common name from the fact that the moths usually emerge and lay their eggs in the fall, mostly in November; often a few, however,

Fig. 88. — Fall canker-worm, female moth (×2¾).

and sometimes many, of the moths do not emerge until spring. The male moths are about the same size and resemble much those of the spring canker-worm, but their wings are of a darker smoky or brownish-gray color; the fore wings

Fig. 89. — Fall canker-worm moths laying eggs. Knight photo (× 2).

Fig. 90.—Portion of an egg-mass of the fall canker-worm, greatly enlarged.

have a distinct whitish spot on the front edge near the tip (Fig. 87). The wingless female moths are nearly $\frac{1}{2}$ an inch in length and of a uniform brownish-ash color (Fig. 88). There are no spines on the bodies of either the male or female fall canker-worm moths. The dark grayish eggs resemble tiny flower-pots or inverted truncated cones, the flattened steel-gray

Fig. 91. — Eggs of the fall canker-worm, side view, greatly enlarged.

Fig. 92. — Fall canker-worm, full-grown caterpillar (× 2).

FIG. 93. — Fall canker-worms
(× 1½).

top marked by a darker ring and central spot (Figs. 90 and 91). They are laid in exposed positions on the bark (Fig. 89), mostly on the twigs, in flattened masses of from 100 to over 400, and are set close together on end in quite regular rows.

The eggs hatch in April or May when the buds are opening, and the caterpillars work on the foliage for about a month in the same manner as the spring canker-worms. The full-grown fall canker-worms are about an inch in length, slender, of a general black color with the venter and all the legs light apple green (Figs. 92 and 93). Just below the spiracles there is a stripe of light lemon yellow, and above them on each side of the body are three narrower whitish stripes. There are 3 pairs of pro-legs, but the first pair are much smaller and are not used when the caterpillar loops or spans in walking. About June 1, most of the caterpillars spin down from the trees and enter the ground from 1 to 4 inches, where they construct a thin but dense and tough silken cocoon in which the greenish-brown, somewhat tender pupæ (Fig. 94) remain until November, or sometimes until the next spring. There is only one brood annually.

FIG. 94. — Pupæ of fall canker-worm. Enlarged.

Natural enemies.

The eggs are sometimes devoured by a mite, *Nothrus ovivorus*, or a minute parasitic chalcid fly may develop in them. The chickadees often find many of the eggs or the egg-laden female moths on the bark. The caterpillars are attacked by a Tachina fly, and the same ground beetles and birds that devour the caterpillars of the spring canker-worm also include this species in their menu.

Remedial measures.

The remedial treatments for the fall canker-worm are the same as recommended for the spring canker-worm, but the sticky band or other barrier for preventing the ascent of the wingless female moths must be applied in the fall, in October, and kept in working condition until December, then renewed in February or March to get the belated ones that may not emerge until spring.

Further details regarding the habits, natural enemies and remedial treatments for this species, will be found in the preceding more extended account of the spring canker-worm.

The Lime-tree Span-worm

Erannis tiliaria Harris

This native American insect is widely distributed and common in apple orchards, and on basswood, elm and other forest trees. It often works with both the spring and fall canker-worms in orchards, but is rarely so numerous and injurious. The moths appear in November and are active at night only. The practically wingless, yellowish-white females, nearly $\frac{1}{2}$ an inch long, are marked with two rows of black spots down the back (Fig. 95). The light, rusty buff-colored front wings of the male moth expand about $1\frac{1}{2}$ inches and are crossed by two narrow, wavy, darker stripes; the hind wings are much lighter

FIG. 95. — The lime-tree span-worm, female moth (× 3½).

(Fig. 96). The females crawl up the trees and lay their oval, light greenish-yellow, finely pitted eggs (Fig. 97) from 1 to 5 in a place tucked away out of sight beneath the scaly bark or in crevices on the trunk and larger limbs. The eggs hatch in early spring, often in April, and some of the caterpillars continue feeding until nearly the middle of June. When full-grown they are nearly 1½ inches long and the skin has a rough velvety appearance. The head is rusty red in color and much roughened. Along the back extend 10 narrow, crinkly, black stripes separated by similar light yellow stripes, the yellow rarely predominating. A broad lemon yellow stigmal stripe extends along each side, and the underside of the body, including the two pairs of pro-legs, is light yellowish-white in color (Fig. 98).

FIG. 96. — The lime-tree span-worm, male moth.

FIG. 97. — Eggs of the lime-tree span-worm. Enlarged.

Fig. 98. — Three views of the lime-tree span-worm caterpillar ($\times 1\frac{1}{2}$).

Early in June the caterpillars are quite restless, and finally go into the ground an inch or more, where they transform to brown pupæ (Fig. 99) in simple, earthen cells. Some of them pupate in May. There is but a single generation annually, the pupa stage continuing until November.

Birds, especially the bluejay, get many of the caterpillars. The fiery and rummaging ground-beetles (*Calosoma calidum* and *scrutator*) also climb the infested trees and capture them.

Remedial treatment.

As its habits and life history are similar to those of the fall canker-worm, this lime-tree measuring-worm can be controlled by the same barrier and spraying methods. About a week before they go into the ground to transform, these large caterpillars are very restless and often either drop or are blown from the trees to the ground. In their wanderings and attempts to get on to the trees again, many have been caught in the wire-screen barriers applied earlier to prevent the ascent of the spring canker-worm moths.

Fig. 99. — Pupa of the lime-tree span-worm. Enlarged.

The Mottled Umber-moth

Erannis defoliaria Clerck

This common and destructive European orchard pest has established itself in British Columbia, and since 1893 it has been more or less destructive in plum and cherry orchards. The caterpillars are general feeders on various fruit and shade trees and often gnaw into unripe cherries in England. The life history and habits of this mottled umber-moth are very similar to those of the American lime-tree span-worm, the moths emerging in November and the caterpillars working in June and early July in British Columbia.

The European insect differs but little from the native species. The wingless females are brownish with rows of brown instead of black spots, and the dull ochre-brown front wings of the males are crossed by two wider dark waved bands, while the pale hind wings as well as the front ones are mottled with brown dots. Descriptions and figures of the caterpillars indicate that they are much like the lime-tree span-worms; the dorsal region is described as reddish-brown instead of yellow between the black stripes and the spiracles are in the center of blotches of reddish-brown.

A Tachina fly parasite is killing some of the caterpillars in British Columbia. This imported pest will doubtless succumb to the same remedial treatment as the native species.

Bruce's Measuring-worm

Rachela bruceata Hulst

This insect was first described from western New York in 1886, where a few years later it was quite injurious in apple orchards and on maple trees. Millions of the caterpillars appeared in Alberta, Canada, in 1902, and the following year hundreds of acres of American aspen or poplar trees were denuded in June. The moths have been found in British Columbia also.

The wingless female moths are about one third of an inch long, light brownish-gray, and closely resemble the females of the fall canker-worm, differing in being only about two thirds as large

FIG. 100. — Bruce's measuring-worm, female moth (× 4).

and in having slightly longer stubs of wings (Fig. 100). The winged male moths have a wing expanse of about $1\frac{1}{8}$ inches and are of a general very pale brownish color, the wing veins being quite distinctly outlined by darker scales (Fig. 101). The moths emerge and go up the trees for egg-laying at night in October and November. The reddish-orange, oval, finely pitted

FIG. 101. — Bruce's measuring-worm, male moth (× 1½).

FIG. 102. — Eggs of Bruce's measuring-worm. Enlarged.

eggs (Fig. 102) are usually laid singly where readily seen in the crevices of bark scales on the trunk and larger branches (Fig. 103). The eggs hatch in April, or as soon as the buds begin to open.

FIG. 103. — Eggs of Bruce's measuring-worm tucked behind a bit of lichen on a tree trunk. Enlarged.

The young canker-worms often bore into the buds and continue feeding for 4 or 5 weeks. They seem to be particularly fond of the opening flower buds on apple, and thus often ruin a prospective crop. The full-grown Bruce's canker-worms are about ¾ of an inch long and of a general apple green color, with three narrow yellowish-white stripes along each side of the body (Fig. 104). There are two pairs of pro-legs, and the head, thoracic and anal shields, and a large spot on the anal pro-legs, are usually blackish, but sometimes nearly concolorous with the body in lighter colored specimens.

Fig. 104. — Bruce's measuring-worm, full-grown caterpillars, slightly enlarged.

By June 1, and sometimes a month before in earlier seasons, these canker-worms all enter the ground a short distance, and in a slight silk-lined cocoon (Fig. 105) they soon transform to light brown pupæ (Fig. 104) from which the moths do not emerge until autumn. Like the other kinds of measuring-worms known as canker-worms, there is thus but a single generation of this species annually. It ap-

Fig. 104 *a.* — Pupa of Bruce's measuring-worm. Enlarged.

Fig. 105. — Cocoon of Bruce's measuring-worm.

parently gets through with its destructive work a little earlier in the spring than the other canker-worms, and evinces a special fondness for the blossom buds. It can be controlled by the remedial measures recommended for the fall canker-worms.

The Half-winged Geometer

Phigalia titea Cramer

About 1898 several kinds of canker-worms devastated apple orchards in western New York, and among them was this species, which we have designated the half-winged canker-worm, because of the nearly half developed condition of the wings of the female moths. The insect is widely distributed from New York to Minnesota and includes among its food-plants, apple, blackberry, rose and several forest trees. It is rarely very numerous,

but the peculiar cater-
pillars and female moths
often attract the atten-
tion of the orchardist.
The male moths are pale
ash gray with 3 blackish-
brown, narrow, trans-
verse stripes across the
front wings which expand
nearly 1½ inches; the an-
tennæ are strongly pec-

FIG. 106. — The half-winged geometer, male
moth (× 1¼).

tinated (Fig. 106). The females are about ⅝ of an inch long,
the hind wings reach to the second abdominal segment, and
the body is clothed with a mixture of black, brown and white
scales, the lighter scales predominating on the wings and
underside (Fig. 107).

The moths emerge and go up the trees at night in March
and April, and one fe-
male may lay 600 eggs.
The slightly egg-shaped
greenish eggs have a firm
shell covered with a net-
work of fine ridges form-
ing 6-sided areas (Fig.
108). The caterpillars
are full-grown about the
middle of June, when
they are about 1½ inches
long, and of a general
violaceous-brown color
with a rough, black-
mottled head. Eight
pairs of narrow, irregular,
black stripes extend along

FIG. 107. — The half-winged geometer, female
moth (× 2½).

H

FIG. 108. — Eggs of the half-winged geometer. Enlarged.

the body (Fig. 109), the four pairs on the underside being less distinct and ending at the first of the two pairs of pro-legs. The hair-bearing spots are elevated into shining black papillæ, those in the hind row on the first four abdominal segments being considerably larger. These canker-worms pupate (Fig. 110) late in June in simple earthen cells in the soil, and the single annual life-cycle is completed when the moths emerge in autumn.

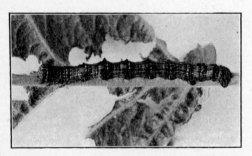

FIG. 109. — Caterpillars of the half-winged geometer ($\times 1\frac{1}{3}$).

FIG. 110. — Pupa of the half-winged geometer. Enlarged.

This half-winged canker-worm will readily succumb to the same remedial treatments as advised for the fall canker-worm.

THE WHITE ENNOMID

Ennomos subsignarius Hübner

This common and widespread measuring-worm often strips various forest trees and shrubs, and it has also defoliated apple orchards in Georgia and Kentucky. The moths are snow-white, with a wing-expanse of about 1½ inches; the males have strongly pectinate antennæ. It is said that if a bird alights in a tree where the moths are numerous, they suddenly drop like snowflakes to the ground for protection. The caterpillars also drop at the slightest jar and swing in the air by their silken threads. They are about 1¾ of an inch long, of a reddish-black color and have two pairs of pro-legs and three pairs of small tubercles on the back. The caterpillars transform through oddly granulated, brown-dotted pupæ to the moths in about 10 days in May or June in one or more leaves rolled or loosely fastened together. The snow-white moths appear mostly in June and lay their pouch-shaped, greenish-olive eggs in large patches of a hundred or more on the undersides of the upper branches of the trees. The eggs are set on their rounded ends, the top being cut off rather squarely and marked with a narrow, white, oval ring surrounding a darker area. There is but one generation annually, and about 9 months are spent in the egg stage, the eggs hibernating and hatching in April or May.

This measuring-worm can be readily controlled in orchards by thoroughly spraying the trees in May when the caterpillars are small with arsenate of lead, 5 pounds in 100 gallons of water.

REFERENCE

Cornell Agr. Exp. Sta. Bull. 286. 1910.

The White-marked Tussock-moth

Hemerocampa leucostigma Smith and Abbot

There are three species of these tussock-moths that may injuriously infest orchards. Two of these are native American insects and one is an old and common European species.

Fig. 111. — Caterpillar of the white-marked tussock-moth, full-grown ($\times 1\frac{1}{3}$).

With their many hairs arranged in striking pencils, tufts and tussocks or brushes, the caterpillars of the white-marked tussock-moth present a very handsome and characteristic appearance (Fig. 111). They are about $1\frac{1}{2}$ inches long when full-grown and of a general dark gray color with a broad velvety black band bordered by yellow stripes on the back and a similar yellow stripe along each side below the spiracles. The head, thoracic shield and two raised glands on the back of the 6th and 7th abdominal segments are bright vermilion-red. Their striking characteristics are dense, brush-like, cream-colored tufts or tussocks of hairs on the back of each of the first four abdominal segments, and pencils of long plume-tipped black hairs projecting from each side of the first thoracic segment and from the back of the eighth abdominal segment. These strikingly beautiful caterpillars are common in orchards, especially on apple, pear, quince and plum trees, in Canada and over the eastern half of the United States. Considerable injury often

FIG. 112. — Pupæ of the white-marked tussock-moth. Enlarged.

results from their work on the foliage in orchards, and in one case 25 per cent of the apples were ruined by the caterpillars gnawing into the sides. But the insect often be-

FIG. 113. — Female white-marked tussock-moths depositing egg-masses on cocoons.

comes a far more destructive and formidable pest in cities and towns on shade trees, especially the horse chestnut, poplar and elm.

In the North there is but a single annual brood, but in southern New York and southward there are two or three broods. The caterpillars feed mostly from the underside of the leaves during June, and in July they transform to pupæ (Fig. 112) in their silken cocoons, in which their long hairs are interwoven, on the bark in the crotches of the trees or on fences or houses near by. In about two weeks the moths emerge. The hairy, grub-like, light-grayish females have mere stubs of wings, and usually remain on their empty cocoons until

after they mate and lay a mass of from 150 to over 700 eggs thereon (Fig. 113). The nearly spherical, yellowish-white eggs are covered by a mass of conspicuous white, frothy material. The ashy-gray colored male moths have feathery antennæ and well developed wings, which expand about 1⅜ inches (Fig. 114). The front wings are crossed by undulated bands of darker shades and bear a conspicuous white spot near the anal angle, hence the name *leucostigma* or white-marked. The winter is always passed in the egg stage, the caterpillars hatching late in May in New York.

Fig. 114. — Male white-marked tussock-moth.

Natural enemies.

This tussock-moth is beset by many enemies. At least 10 different birds eat the caterpillars and doubtless do much to keep the insect in check in orchards and the open country. Several species of shield-bugs and the southern wheel-bug attack the caterpillars and pupæ; the pupæ are also eaten by small red ants. The grubs of two Dermestid beetles and a species of mite may devour the eggs. And as many as 90 per cent of the caterpillars and pupæ sometimes fall a prey to more than 20 different kinds of hymenopterous and dipterous insect parasites, the most effective of these little enemies being *Pimpla inquisitor, Chalcis ovata, Tachina mella, Frontina frenchii,* and *Euphorocera claripennis*. Unfortunately, however, there are 14 hyper-parasites which work on the true parasites and thus materially lessen their effectiveness. There are also tertiary parasites which destroy these hyper-parasites, thus presenting a very complicated and interesting case of insect parasitism.

Remedial measures.

A practicable and effective method of controlling this pest is to collect and burn the eggs in autumn or winter. The grayish egg-masses are quite conspicuous on the bark and they are often attached to a dead leaf or two fastened to the branches. Where shade trees are infested in cities, it will pay to employ laborers to collect the eggs, and sometimes the school children can be induced to do very effective work by offering prizes or by paying liberally for certain quantities of the egg-masses. The latter method was employed successfully several years ago in Rochester, N. Y., during a severe outbreak of the pest.

Before they are half-grown the beautiful caterpillars will succumb to thorough applications of a strong poison spray, such as Paris green (1 pound in 100 gallons) or arsenate of lead (5 or 6 pounds in 100 gallons). One or two applications of such a spray will usually control this pest. Later when the caterpillars are larger, the poison is not so effective. As the caterpillars drop to the ground by a silken thread when the tree is jarred, some orchardists have found it practicable to capture them on curculio-catchers or sheets. Or after jarring them off, they can be prevented from ascending the trees by means of a sticky rope band around the trunk. Treat the rope with a tanglefoot mixture of resin and castor oil, as recommended for cankerworm bands.

REFERENCES

U. S. Bur. Ent. Tech. Bull. 5. 1897. *Parasites.*

U. S. Farmers' Bull. 99, pp. 14–31. 1899.

Del. Agr. Exp. Sta. Bull. 56, pp. 9–18. 1902.

N. Y. (Geneva) Agr. Exp. Sta. Bull. 312. 1909.

The California Tussock-moth

Hemerocampa vetusta Boisduval

This native California tussock caterpillar has been recorded only from that state, where it is common on live oak and yellow lupin trees, and has injuriously infested apple and cherry orchards. There is but a single brood annually, the moths appearing in May, June and July and laying their eggs on their empty cocoons in light grayish masses covered with hairs from the mother's body. Some of the egg-clusters hatch as early as February, while others may not until April or May. The caterpillars feed on the foliage from one and a half to two months. They have black heads, crimson hair-bearing warts and pro-legs, and the four tussocks or brush-like tufts of hairs on the back are often dark gray with brownish crests. The practically wingless grub-like female moths have dark brown bodies covered with sordid white hairs. The males resemble those of the antique tussock-moth but are of a darker chestnut-brown color. The general life-habits of this California tussock-moth are similar to those of the two eastern species.

Natural enemies.

Tachina flies and other parasites sometimes destroy half of the caterpillars and pupæ, but the birds do not seem to eat them to any extent in California. The grubs of a Dermestid beetle devour some of the eggs, and others are parasitized by a minute hymenopterous insect, *Telenomus orgyiæ.*

Remedial treatments.

Poison sprays are reported as not very effective against the caterpillars in California, the older ones after eating poisoned foliage being able to transform to the moths. The measures recommended are the collecting and destroying of the eggs during the winter season, and the beating or jarring of the caterpillars from the trees, then preventing their ascent with sticky rope bands, as described for the white-marked tussock-moth.

REFERENCE

Cal. Agr. Exp. Sta. Bull. 183. 1907.

THE ANTIQUE TUSSOCK-MOTH

Notolophus antiqua Linnæus

This common European pest has been more or less injurious in America on shade trees and in orchards for nearly a century. It is often seen in Nova Scotia, the New England states and also occurs as far westward as Washington. The caterpillars resemble those of the white-marked tussock-moth, but the head

FIG. 115. — The antique tussock-moth, full-grown caterpillar.

is black, and the first two tussocks or brushes of hairs on the back are black on the young caterpillars, but become whitish in later stages (Fig. 115) After the third stage there is also an additional pencil of long, black, plume-tipped hairs projecting from each side of the second abdominal segment.

In its life history this species differs but little from that of the white-marked tussock-moth. There is a single brood annually in its northern range, the eggs hibernate, but there is no covering on the egg-masses laid on the cocoons. In England it is said that a few of the eggs hatch at a time over a period of 10 weeks, but they seem to hatch uniformly in America. The grub-like, practically wingless female moths have blackish bodies covered with yellowish-white hairs. The males have rust-brown colored wings, the front ones being crossed by two deeper brown bands and having a very conspicuous white spot near the anal angle.

This antique tussock-moth is attacked by many of the same enemies as the white-marked tussock-moth, and it can be controlled by the same remedial measures.

THE ORIENTAL MOTH

Cnidocampa flavescens Walker

This is an Asiatic insect which was apparently introduced into Massachusetts on nursery stock from Japan some time before 1900, but its presence was not discovered until February, 1906. One of the peculiar cocoons of this moth has also been found in an Albany, New York greenhouse on imported Japanese maples. It seems to prefer to feed on the foliage of Norway maples, but pear, apple and cherry are often infested, and it includes a dozen other shade and forest trees among its food-plants. The yellowish-red, slug caterpillars are armed with rows of spiny tubercles or horns, and are about ¾ of an inch long. They hibernate in curious tough, smooth, oval-shaped cocoons, a little more than ½ an inch in length, fastened firmly to the bark, mostly on the smaller branches in or near the axil of a branch. The cocoons are strikingly colored, with whitish and brown often so mingled as to suggest certain oriental designs. There is a brown lid at one end which is pushed open by the pupa late in June when the moth emerges. There is but one brood annually, the cocoons being made in September or October. The pretty moths have a wing expanse of nearly 1½ inches, the head, thorax and inner half of the front wings above being of a dull chrome yellow color, while the upper portion of the outer half is light chestnut brown with a yellowish tinge and the lower half is tinged with pink.

While the oriental moth belongs to a family, the Cochlidiidæ, the members of which are not usually of economic importance, its natural enemies probably were not introduced with it into this country, and it is maintaining itself and slowly spreading in Massachusetts. Judging from its range in Asia, the insect can live almost anywhere in the United States as far as climate is concerned. Whether it ever becomes a serious pest in America, especially on fruit trees, only the future can reveal, but the probabilities are that it will not.

The caterpillars will doubtless succumb to a thorough application of a strong poison, like arsenate of lead, 4 pounds in 100 gallons of water, and it would be practicable to prune off and destroy the curious cocoons in winter or early spring on small trees.

REFERENCE

Mass. Agr. Exp. Sta. Bull. 114. 1907.

The Fall Webworm

Hyphantria cunea Drury

The unsightly nests of this caterpillar are familiar objects in late summer on a large number of forest and shade trees (Fig. 116). Economically this insect is of greater importance as a shade tree pest in cities than as an enemy of fruit trees, but occasionally its attacks on apple and pear are sufficiently severe to cause considerable loss. It feeds on over one hundred different trees and is distributed over the Eastern states from Georgia and Texas to Montana and Canada; it also occurs in California. In the North, where only one generation develops annually, it only occasionally causes serious injury to fruit trees, but from southern New York southward, where there are two full generations, orchards are sometimes completely stripped of their leaves. In the South it sometimes becomes a serious enemy of the pecan.

The winter is passed by the reddish-brown pupæ, $\frac{1}{2}$ inch in length, in thin cocoons placed in crevices of the bark, under trash, or just beneath the surface of the ground. The emergence of the moths extends over a considerable period, from early June until the middle of August in New York. The webs begin to become conspicuous in July and are most abundant in August (Fig. 117). In the District of Columbia the first brood of moths appears soon after the leaves are fully developed; that is, about June first, the second brood occurring in July and August.

Fig. 116. — Fall webworm, beginning of a nest.

The moths expand about $1\frac{1}{4}$ inches and vary in color from satiny white to white heavily spotted with black or brownish spots. Some of these forms have received specific names, but it is now generally believed that we have only one species which is highly variable. More recently Dr. H. H. Lyman has attempted to separate two forms on the color of the larvæ, but the evidence presented is hardly convincing. More extensive rearings are greatly needed.

The moth lays her pale green eggs in clusters on the upper or under surface of the leaves, and covers the cluster with white hairs from her body (Fig. 118). The cluster is about $\frac{3}{8}$ inch

in diameter and
contains 300 or 400
eggs. The egg is
globular, $\frac{1}{50}$ inch in
diameter, and has
a delicate thimble-
like sculpture. The
eggs of the first
brood hatch in
about 10 days and
those of the second
in about a week
after deposition.

Soon after hatch-
ing the extremely
hairy young cater-
pillars begin feeding

Fig. 117. — Fall webworm, nest.

on the leaves, usually at the end of a branch, which they
inclose in a silken web. At first they feed only under cover
of the web, which is enlarged to inclose more foliage as
there is need of fresh food. When
the caterpillars become larger they
leave the nest at night and feed in
the open. As food becomes scarce
on a branch, all or part of a colony
may migrate to another branch,
where a new nest is formed, or in
case the whole tree has been defoliated, they may crawl to another
tree.

Fig. 118. — The fall webworm,
moth laying eggs.

The full-grown caterpillars (Fig.
119) are about $1\frac{1}{4}$ inches in length
with a broad dusky stripe along the back and usually with
yellowish sides thickly spotted with small blackish dots. A

Fig. 119. — Full-grown fall webworm caterpillars. Slightly enlarged.

darker form occurs in which the yellow on the sides is entirely obscured, and the whole caterpillar is of a dull dusky color, but all gradations are found.

Natural enemies.

Few birds feed on these hairy caterpillars to any great extent. Among predaceous insects their most important enemies are the Mantis (*Mantis carolina*) and the wheel-bug (*Prionidus cristatus*), both abundant in the South. Of true parasites the most important are *Telenomus bifidus* Riley, attacking the eggs, *Meteorus hyphantriæ* Riley, *Apanteles hyphantriæ* Riley, and *Limneria pallipes* Prov., attacking the caterpillars. Good work is also done by an unnamed Tachinid-fly. Unfortunately the efficiency of this series of parasites is greatly lessened by the great number of secondary parasites, which prey upon them. Under favorable conditions great numbers of the caterpillars are killed by a fungous disease, *Empusa grylli*.

Remedial measures.

In the forest we must depend on its natural enemies to hold this pest in check. On shade trees in cities spraying with an arsenical poison and a thorough and timely removal of the small webs while the caterpillars are still young will do much to prevent serious damage. Especial attention should be given to the destruction of the first generation, thereby lessening the difficulty of controlling the second. In orchards where arsenical spraying is regularly practiced most of the first brood of caterpillars will be destroyed. The webs should be cut out or burned out with a torch as soon as observed. This can be done most easily while they are small and before any large branches have been included in the nest. The webs should be destroyed during the day while practically all the caterpillars are within.

REFERENCES

Riley, Rept. U. S. Com. Agr. 1886, pp. 518–539.
Lyman, 32d Rept. Ent. Soc. Ont., pp. 57–62. 1902.

Del. Agr. Exp. Sta. Bull. 56. 1902.
Felt, Forest Insects, I, pp. 142–146. 1905.
Berger, U. S. Bur. Ent. Bull. 60, pp. 41–51. 1906.

THE APPLE-TREE TENT-CATERPILLAR

Malacosoma americana Fabricius

The unsightly nests (Fig. 120) of this caterpillar are conspicuous objects on wild cherry trees and in neglected orchards.

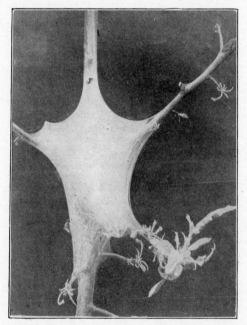

FIG. 120. — The nest of the apple-tree tent-caterpillar.

This species occurs in the Eastern states and Canada, ranging westward to the Rocky Mountains, where it is replaced by other forms. It occurs sparingly in California. Its favorite food is the wild cherry, although it will attack apple, peach, plum, and more rarely, witch-hazel, beech, birch, barberry, oak, willow and poplar. When excessively abundant apple trees are frequently completely defoliated and killed (Fig. 121). Destructive outbreaks usually continue for two or three years

Fig. 121. — Tree defoliated by apple-tree tent-caterpillars.

only and are then followed by a longer period during which the species is rarely noticed. This periodic fluctuation is thought to be the result of complicated interrelations existing between the caterpillar and its parasites, and is a striking illustration of the ups and downs of insect life.

I

FIG. 122. — Egg-mass of the apple-tree tent-caterpillar cut open to show eggs.

The winter is spent in the egg stage, although the embryo is fully developed in the fall. The eggs are elongate, thimble-shaped, about $\frac{1}{25}$ inch long and are laid in masses of over 300 to 400, usually encircling a small branch as a broad band (Fig. 122). The whole egg-mass is covered by a brownish gluey froth, which protects it from the weather. About the time the first buds open the eggs hatch and the young larvæ begin to feed on the opening leaves. The larvæ are

FIG. 124. — Cocoons of the apple-tree tent-caterpillar ($\times \frac{5}{8}$).

FIG. 123. — Full-grown apple-tree tent-caterpillars ($\times 1\frac{1}{9}$).

social, and all those from a single egg-cluster remain together and soon begin a silken nest. Sometimes when two egg-masses are placed close together the two colonies unite in forming a common nest. The nest is at first small, but gradually enlarges as the caterpillars grow larger, until in some cases it may be nearly two feet in length.

During storms and the heat of the day the caterpillars usually remain within the nest, coming out to feed early in the morning, in the

FIG. 125. — Pupæ of the apple-tree tent-caterpillar (× 3).

evening, or at night when it is not too cold. The full-grown caterpillar is about two inches long, black with a light stripe down the back and with dots of blue and white along the sides, and is clothed with fine, soft, yellowish hairs (Fig. 123). When nearly full-grown they wander from the nest, and after feeding for a few days more crawl to some protected place and spin their cocoons.

The oval, white cocoons are about one inch in length (Fig. 124). They are made of tough, closely woven, white silk, and are held in place by a few irregular coarser threads. The

FIG. 126. — Male moths of the apple-tree tent-caterpillar.

newly made cocoon is dusted over with a yellowish powder.
A few days after finishing the cocoon the larva transforms into
a brownish pupa, $\frac{5}{8}$ to $\frac{3}{4}$ inch in length, which is clothed with a
brownish pubescence except on the sheaths of wings and legs
(Fig. 125).   The pupal period lasts about three weeks.

Fig. 127. — Male and female moths of the apple-tree tent-caterpillar.

In New York the majority of the moths emerge during the
last week of June and the first week of July, and the eggs are
laid soon after.   There is but one generation a year, from
July till April being spent in the egg stage.   The moths are
dull, reddish-brown, marked on the front wings by two nearly
parallel oblique whitish lines (Figs. 126 and 127).   The females
expand $1\frac{1}{2}$ to 2 inches; the males are slightly smaller and of a
darker brown.

Natural enemies.

The tent caterpillar is held in check by a long series of parasites which prey upon it in the egg, larval and pupal stages. This control is so effective that for many years the insect is rarely noticed. Then the time comes when, owing to the rarity of the host, the parasites are reduced in numbers to a still greater degree, and the caterpillars again have an opportunity to multiply unchecked. The resultant outbreak may be very serious, as was the case in New York and New England in 1897 and 1899. Four species of parasites have been reared from the egg and over 20 from the larva and pupa. Birds and toads also feed on the caterpillars to a considerable extent, and many nearly full-grown larvæ die of a bacterial disease.

Means of control.

In orchards which are well sprayed as for the codlin-moth and curculio, tent-caterpillars are rarely troublesome. The young larvæ are readily poisoned by either Paris green or arsenate of lead. On peach and plum, which are not commonly sprayed with an arsenical, it will pay to keep careful watch for the conspicuous egg-rings while pruning. They should be removed and burned.

The nests may be destroyed by wiping out with the hands. It should be done while the caterpillars are at home, which is usually the case during stormy weather and in the heat of the day. Burning out the nests is not to be recommended as there is danger of injuring the tree; burned areas in the bark often develop cankers that may destroy the whole branch.

The wild cherry is the favorite food-plant of this insect. When growing along roadsides and fences and in other waste places these trees are usually worthless and should be cut down, as they serve as centers of infestation for near-by orchards.

REFERENCES

N. Y. (Geneva) Agr. Exp. Sta. Bull. 152, pp. 279–297.   1898.
Conn. Agr. Exp. Sta. Bull. 139.   1902.
N. H. Agr. Exp. Sta. Tech. Bull. 6.   1903.   *Parasites.*

The Western Tent-caterpillar

Malacosoma fragilis Stretch

From the Rocky Mountains westward the apple-tree tent-caterpillar is replaced by a number of forms all very much alike in habits and in the form and coloration of the moths. The caterpillars, however, are different. From the Rockies to the Sierras and from Mexico to Canada the dominant form is

Fig. 128. — Western tent-caterpillars.

M. fragilis. It forms a web like its eastern relative and has a similar life history.

In Colorado the larvæ become full-grown by July 1, and the moths emerge and lay their eggs during the latter part of the month. The full-grown larva is similar to the eastern form but is pale blue or blue-gray on the sides, and the median line has a row of bluish spots instead of the whitish line (Fig. 128). The species is single brooded. The larvæ feed on apple, willow, poplar, wild cherry, wild rose and wild gooseberry.

Control.

The same remedial measures are suggested for this species as for the apple-tree tent-caterpillar.

Another species, *Malacosoma pluvialis* Dyar occurs in the Pacific Northwest. Very little is known concerning it except that in habits and life history it does not differ greatly from the other species.

Reference

Ore. Agr. Exp. Sta. Bull. 33. 1894.

The Forest Tent-caterpillar

Malacosoma disstria Hübner

The common name of this species is a misnomer, for the caterpillars do not construct a true tent, as in the case of the preceding species. It is closely related to the apple-tree tent-caterpillar, and has a similar life history, but its habits are decidedly different. Its range extends throughout the United States and Canada and it has been reported from Mexico. It is naturally a forest insect, the maple being its favorite food plant. When unusually abundant, however, it attacks a large number of trees and shrubs, and in the orchard feeds on the apple, plum, peach, cherry and pear. Severe outbreaks occur at

Fig. 129. — Egg-rings of the forest tent-caterpillar.

rather long intervals and usually last for only two or three years. In 1898 and 1899 a very extensive and severe outbreak occurred in New York and New England; the caterpillars defoliated thousands of acres of forest and did great damage in fruit orchards. They also appeared in destructive numbers in 1912 and 1913.

The winter is passed in the egg stage (Fig. 129). The eggs hatch in early spring and the young caterpillars feed in colonies on the opening leaves. They leave a silken thread wherever they go and in this way the colony frequently webs in some of the leaves where they are feeding. These slight webs are not true tents; they are never entered by the caterpillars and are not used for protection. When not feeding or when preparing for molting the larvæ congregate in masses on the branches or trunk (Fig. 131).

When the caterpillars become nearly full-grown they become restless and wander away to feed singly for some days. The full-grown caterpillar is about two inches long and is easily distinguished from the apple-tree tent-caterpillar by having a median row of wedge-shaped or club-shaped cream-colored spots instead of the continuous median whitish stripe present in that species (Fig. 130). On the abdominal segments these spots are broken into two unequal parts, which gives them the appearance of a row of exclamation marks. In New York the caterpillars become full-grown in late May and early June, depending on the season.

Fig. 130. — Full-grown forest tent-caterpillars ($\times 1\frac{1}{3}$).

The cocoons are made of white silk, the outer layers being

soft and fluffy and dusted with a yellowish powder (Fig. 132). Most of the cocoons are placed in curled leaves, but many are found in other more or less protected situations. In New York the moths emerge and lay their eggs the last week of June and the first of July. The moths are similar to those of the apple-tree tent-caterpillar, but the oblique bands across the front wings are brown instead of whitish (Fig. 133). The eggs closely resemble those of that species, but the egg-rings are smaller and more abruptly rounded off at the ends. Each egg-mass contains from 150 to over 400 eggs. The young caterpillar becomes fully formed by the end of August but remains within the egg until the following spring.

FIG. 131. — A mass of forest tent-caterpillars resting on a tree trunk.

Natural enemies.

The caterpillars are preyed upon by several kinds of birds and by toads. Two predaceous beetles, *Calosoma scrutator* Fab. and *C. calidum* Fab., feed on the larvæ as do two of the sucking bugs, *Podisus placidus* Uhler and *P. serieventris* Uhler. A number of hymenopterous and dipterous parasites prey on the caterpillars and are valuable aids in holding the pest in check. The most important of these are, *Pimpla conquisitor* Say, *Pimpla inquisitor, Anomolon exile* Prov., *Tachina mella* Walk. A mite also destroys the eggs.

Fig. 132. — Cocoons of the forest tent-caterpillar.

Methods of control.

In the forest little can be done to fight this insect. As a shade tree pest in villages and cities spraying with arsenate of lead, 8 pounds to 100 gallons of water, while the larvæ are still small, would be effective. The cocoons should be collected whenever found and placed in a box covered with $\frac{3}{16}$ inch mesh screen, which will permit the escape of the parasites while retaining the moths. In some towns children have been hired to collect the cocoons at ten cents a quart with excellent results.

Fig. 133. — Female moth of the forest tent-caterpillar.

In orchards where arsenical spraying is practiced, as for the codlin-moth, the young caterpillars are killed before doing much injury. On small trees they may be jarred off on to sheets or curculio-catchers and destroyed. When molting, and during the heat of the day, the caterpillars collect in large masses on the trunk and branches, where they may be readily brushed down and crushed. After defoliating a piece of woodland the caterpillars frequently migrate to near-by orchards in countless numbers. Banding the trunks with tree tangle foot or loose cotton bands will prevent the ascent of these wandering larvæ. The egg-rings are easily seen while pruning, especially on small trees. They should be removed and burned, for if thrown on the ground the young larvæ may be able to crawl to the tree and so survive.

REFERENCES

N. Y. (Geneva) Agr. Exp. Sta. Bull. 159. 1899.

THE YELLOW-NECKED APPLE CATERPILLAR

Datana ministra Drury

In the Northern states and Canada apple branches are often defoliated in late summer by colonies of black and yellow striped caterpillars about two inches in length when mature; the head is black and the next segment is yellow, whence the common name. While a few colonies are found every year they are only occasionally abundant enough to do serious damage. Besides the apple, this caterpillar also attacks pear, cherry and quince, as well as many forest trees.

FIG. 134. — Moth of the yellow-necked apple caterpillar. Nat. size.

FIG. 135. — Yellow-necked apple caterpillars in resting position.

The moths appear in June and July in New York; they have an expanse of from 1¾ to 2 inches; the front wings are cinnamon-brown, crossed by three or four distinct darker lines; the hind wings are pale straw-color, and the front of the thorax is a rich reddish-brown (Fig. 134). The female deposits her white, ovoid eggs in a flat cluster of from 25 to 100 on the underside of the leaves. The young caterpillars are chestnut-brown in color with obscure darker stripes. At first they feed entirely on the underside of the leaves, but after the second molt they begin eating the edge of the leaves. The segment behind the head now becomes orange or yellow and the body is distinctly striped with black and yellow

FIG. 136. — Cluster of yellow-necked apple caterpillars.

and sparsely clothed with rather long whitish hairs. The caterpillars always feed in colonies and soon strip a branch of its leaves. When disturbed they have the curious habit of bending back the front part of the body with a jerky motion and at the same time elevating the hind end of the body as shown in Figure 135. They become full-grown in August and September, and enter the ground for a few inches and there transform to brownish pupæ (Fig. 137) a little less than an inch in length, without forming cocoons. There is only one brood annually.

FIG. 137. — Pupa of the yellow - necked apple caterpillar. Enlarged.

Control.

The work of these caterpillars is so conspicuous that it is easy to locate them. On small trees it is perfectly practicable to shake them off and crush them on the ground. On larger trees the young caterpillars can be killed by spraying with arsenate of lead, 4 or 5 pounds in 100 gallons of water.

REFERENCES

Beutenmüller, Can. Ent. XX, pp. 16–17. 1888.
N. H. Agr. Exp. Sta. Bull. 139, pp. 213–215. 1908.

THE RED-HUMPED APPLE CATERPILLAR

Schizura concinna Smith and Abbot

Feeding in colonies at the ends of the branches like the preceding species the red-humped apple caterpillar often attracts attention in August and September. It is rarely a serious pest on older trees but the caterpillars sometimes defoliate young trees in August and thus prevent the proper ripening of the wood. It attacks apple, cherry, plum, apricot, pear, blackberry and a number of forest trees.

In the Northern states the inconspicuous grayish-brown

moths fly in June and July. The female has an expanse of about $1\frac{3}{8}$ inches; the male is a little smaller and more distinctly marked. The female deposits her white, nearly round, slightly

FIG. 138. — Moth of the red-humped apple caterpillar. Nat. size.

FIG. 139. — Egg-mass of the red-humped apple caterpillar. Enlarged.

flattened eggs in clusters of 40 to 100 on the underside of the leaves (Fig. 139). The young caterpillars feed at first on the underside of the leaves only, but as they grow larger eat the edges of the leaves. They feed in colonies and are soon able to strip a branch of its leaves. When full-grown the caterpillars are an inch or more in length; the head is coral red, the body is striped with black

FIG. 140. — Red-humped apple caterpillars feeding.

and yellow or whitish lines and on the fourth segment there is a prominent reddish hump (Fig. 140). The body is ornamented with rows of blunt black tubercles, largest on the hump. When at rest the tip of the body is held in an elevated position.

In the North, there is only one brood, the caterpillars maturing in August and September. They construct slight cocoons under trash on the ground and as a rule remain in the larval condition through the winter, pupating the following May or June. In the South where there are two broods, the first brood of caterpillars may form their cocoons in curled leaves.

The red-humped apple caterpillar may be controlled by the measures recommended for the yellow-necked apple caterpillar (page 125).

REFERENCES

Mass. (Hatch) Agr. Exp. Sta. Bull. 28, pp. 17–19.   1895.
Packard, Nat. Ac. Sci., VII, pp. 212–217.   1895.

THE SADDLED PROMINENT

Heterocampa guttivitta Walker

While generally distributed throughout the northern and eastern United States this beautiful green saddle-marked caterpillar had never attracted attention by its injuries until the outbreak of 1907–1908 in New York, Vermont, New Hampshire and Maine, where large areas of forests were defoliated. It is primarily a forest insect feeding on the beech, birch, maple and oak, but is also destructive to the apple.

In Maine the parent moths emerge the latter part of May and during June. The moth has an expanse of about two inches; it is brownish-gray in color and the front wings are crossed by indistinct darker lines. The female deposits her smooth, pale green, slightly flattened eggs singly on the leaves. They hatch in about nine days. The young caterpillars in the first stage bear nine pairs of black horns along the back; the first pair back of the head are much longer than the others and branched like antlers, whence the name antlered maple caterpillar, sometimes given to this species. At first they merely skeletonize the leaves, but after the first molt, feed at the edge of the leaf, eating out portions between the larger veins. The full-grown caterpillar is about $1\frac{1}{2}$ inches in length, and varies greatly in coloration. They are usually pale green and have a conspicuous saddle-shaped mark on the third to the fourth abdominal segments. The tip of the abdomen tapers to a point and is usually held in an elevated position. The caterpillars do not cling tightly to the branches and may be readily shaken to the ground. They become full-grown in about five weeks, having molted four times; they then descend to the ground, where, just below the surface or under the leaf mold, they construct a slight cocoon of silk, within which they change to pupæ, remaining in this condition until the following spring. There is but one brood annually in Maine.

Control.

In the orchard the saddled prominent may be easily destroyed by spraying the trees with arsenate of lead, 4 pounds in 100 gallons of water, soon after the eggs hatch. In case the orchard adjoins woodland or other untreated trees, it may be found advisable to band the trunks with some sticky material like " tree tanglefoot " to prevent the ascent of wandering caterpillars.

REFERENCES

Maine Agr. Exp. Sta. Bull. 161. 1908.
N. H. Agr. Exp. Sta. 19th & 20th Repts., pp. 514–531. 1908.

THE GIPSY MOTH

Porthetria dispar Linnæus

The gipsy moth is a native of Europe, Asia and northern Africa, where it has long been recognized as a serious enemy of orchard and forest trees. There its outbreaks are periodic

and often very severe, vast areas of forests are sometimes
devastated and many park and orchard trees either killed or
seriously injured. It was introduced into Massachusetts in
1869 at Medford near Boston by a French naturalist who was
conducting experi-
ments with silk-
worms. Some of the
insects accidentally
escaped and became
established in the
immediate vicinity
but did not attract
particular attention
for about twenty
years. In 1889,
however, the cater-
pillars appeared in
enormous numbers,
defoliated many
forest, shade and
orchard trees, and
excited great alarm
among the residents
of the region. At
first the state at-
tempted to exter-
minate the pest, ex-
pending large sums

FIG. 141. — Gipsy moth egg-masses in a cavity in
tree trunk.

of money for that purpose, but in 1900 abandoned the project and left the gipsy moth to breed and spread unmo-
lested until 1905, when the enormous losses inflicted and
the continued increase in the size of the infested area com-
pelled the resumption of repressive work. In spite of the
expenditure of immense sums of money by both the state

K

and federal governments, the gipsy moth has gradually extended its range over eastern Massachusetts, Rhode Island, the southeastern part of New Hampshire and has invaded southern Maine. Isolated colonies have also been found in Connecticut, western Massachusetts and New York. It has a wide range of food-plants, including most forest and fruit trees,

Fig. 142. — Full and partly grown gipsy moth caterpillars.

with the exception of ash, juniper and red cedar, and the maples are rarely attacked when other food is available. The caterpillars seem to prefer oak, willow and apple, but will eat almost any kind of foliage when driven to it.

The winter is spent in the egg state. The egg-masses are roughly oval in outline, about an inch in length, light brown in color and covered with hairs from the moth's body. They are attached to the trunk or branches of trees, or are placed in cavities in the tree (Fig. 141), in piles of cord wood, lumber piles, stone walls and stone piles, or in any conveniently sheltered place. Each mass contains normally 400 to 500 eggs, but in cases where the larvæ have been starved they are small and sometimes contain only 50 to 75 eggs. The eggs hatch in the spring just as the buds are bursting and the young, reddish-brown caterpillars feed on the tender leaves, which they riddle with small holes. As they grow larger they devour the whole leaf with the exception of the larger veins. Until about half grown they are able to suspend themselves by a thread of silk and are thus likely to fall on passing vehicles and be transported considerable distances. The larger caterpillars avoid the sun

FIG. 143. — Gipsy moth depositing egg-mass, and pupæ in their cocoons.

as much as possible, feeding at night or in cloudy weather. They become full-grown in about seven weeks or about the first week in July. The full-grown caterpillar (Fig. 142) averages about two inches in length; the ground color is dark gray and there are eleven pairs of prominent tubercles on the back, the first five pairs are blue, the last six dark red. When full-grown or sooner, if the food supply gives out, the caterpillars crawl to some sheltered spot, where they spin a frail

FIG. 144. — Male and female gipsy moths.

cocoon consisting of a few threads of silk, and there transform to dark reddish-brown pupæ (Fig. 143), bearing groups of yellowish hairs. The pupal stage lasts from a week to 17 days.

As a rule, the male moths emerge a little earlier than the females. The male has a light brown body and the wings are yellowish-brown; the front wing is traversed by four wavy dark brown lines. In the female the body is light buff and the wings are grayish white; the dark brown markings on the front wings are similar to those of the male (Fig. 144). The male moth has an expanse of one and a half to two inches, and the females average somewhat larger. The male flies with a characteristic zigzag motion but the female is unable to use her wings and usually deposits her eggs

FIG. 145. — *Calosoma sycophanta*, a European ground-beetle introduced into New England to control the gipsy moth. Enlarged.

within a few inches of the pupal case from which she emerged (Fig. 143).

In its native home the gipsy moth is held in check by its natural insect enemies, both parasitic and predaceous. For several years now the United States Bureau of Entomology has been importing and liberating thousands of these beneficial forms with the hope that ultimately they may be able to hold

FIG. 146. — Combing sticky bands and brushing down caterpillars of the gipsy and brown-tail moths.

the species in check. Some of these introduced insects have already become established in the infested territory and are multiplying rapidly. One of the most promising of these is a large green ground beetle (*Calosoma sycophanta* L.) shown in Figure 145. In both the adult and larval state it preys on the larvæ and pupæ of the gipsy moth.

Means of control.

The control of the gipsy moth in orchards is not a difficult matter. During the winter the trees should be carefully examined for egg-masses and the eggs killed by saturating them with crude coal-tar creosote to which a little lampblack has been added as a marker. The work of destroying the eggs can be greatly facilitated by having the trees properly pruned, by removing all flakes of rough bark and by filling all cavities with cement or covering them with tin. The trees should be sprayed, soon after the eggs have hatched, with arsenate of lead, 10 pounds in 100 gallons of water. If the caterpillars are half-

Fig. 147. — Orchard defoliated by gipsy moth caterpillars, in July.

grown, it is advisable to use 13 or 15 pounds. It is very difficult to poison nearly full-grown caterpillars, and for the best results the spraying should be done while the caterpillars are small. If the orchard is located near untreated woods or other orchards the trees should be banded with tanglefoot to prevent the ascent of caterpillars migrating from the untreated area when the food supply runs short (Fig. 146).

REFERENCES

Forbush and Fernald, The Gipsy Moth, Boston, 1896.
U. S. Dept. Agr. Farmers' Bull. 275. 1907.

U. S. Bur. Ent. Bull. 87. 1910. (Contains references to the important reports on the gipsy and brown-tail moths published by the various New England States.)
U. S. Farmers' Bull. 564. 1914.

The Brown-tail Moth

Euproctis chrysorrhœa Linnæus

This well-known and destructive European caterpillar was accidentally introduced into Massachusetts in the vicinity of

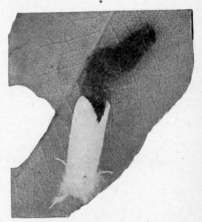

Boston probably in the early nineties, but did not attract attention by its ravages till 1897. Since that time it has gradually extended its ravages so that now the infested area includes part of Rhode Island, the greater part of Massachusetts, southern New Hampshire, southern Maine and extends into New Brunswick and Nova Scotia. The brown-tail moth has been able to spread more rapidly than the gipsy moth, owing to the fact that the

Fig. 148. — Brown-tail moth depositing egg-mass on a leaf.

females are good fliers, and when aided by favorable winds may be transported to a considerable distance. Unlike the gipsy moth, the brown-tail caterpillars do not feed on coniferous trees, their favorite food-plants being apple, pear and oak. The caterpillars, and to a less extent the moths, are provided with minute barbed hairs, which are poisonous to the human skin, causing an annoying and sometimes serious irritation known as the brown-tail rash.

The brown-tail moth has an expanse of about one and one half inches; the wings are white and the tip of the abdomen bears a tuft of yellowish-brown hairs, hence the name of the insect. The males are a little smaller than the females, and the brownish tuft is not so conspicuous. The moths appear the first week in July, and in badly infested areas are often attracted to electric lights in countless numbers. After mating the female moth deposits from 200 to 300 globular yellowish

Fig. 149. — Brown-tail moth caterpillars, one in process of molting.

eggs in an elongate mass on the underside of a leaf (Fig. 148). This egg-mass is about $\frac{3}{4}$ of an inch in length and is thickly covered with brownish hairs from the abdomen of the female. The eggs are deposited during the first three weeks in July and hatch in 15 to 20 days. The young caterpillars feed in colonies on the tender, terminal leaves, webbing them together with silk to form a snug nest (Fig. 150) two or three inches in length within which they pass the winter in a partially grown condition, having molted two or three times before going into hibernation. These winter nests, usually situated at the tips of the branches, are conspicuous objects while the trees are bare of foliage. In early spring, just as the buds are bursting, the caterpillars leave their winter quarters and resume feeding on the unfolding leaves, and if abundant may keep the trees stripped of foliage. They molt four or five times in the spring and become mature toward the last of June. The full-grown caterpillar (Fig. 149) is about 1½ inches in length, nearly black in ground color, clothed with tufts of brownish barbed hairs and has a row of nearly white tufts on each side of the body; there is a coral-red tubercle on the dorsum of the 11th and 12th segments. When mature the caterpillars spin loosely woven cocoons in curled leaves,

Fig. 150. — Winter nest of the brown-tail moth.

crevices in bark of trees, or under any convenient shelter; they are usually found in masses. The pupæ are about ⅝ inch in length and dark brown in color. The pupal period averages about 20 days.

Control.

As an orchard pest, the brown-tail moth can be most readily controlled by collecting and burning the conspicuous hibernating nests during the winter months. The newly-hatched caterpillars can be killed the first or second week in August by a thorough application of arsenate of lead, 8 pounds in 100 gallons of water. Attempts to poison the over-wintering caterpillars, when they appear on the buds in the spring, are not so successful. The caterpillars often devour the leaves as fast as they appear and it is difficult to keep the surface of the rapidly expanding leaves covered with the poison; furthermore, the caterpillars are larger then and consequently harder to kill. If for any reason the destruction of the winter nests has been neglected, and spring spraying must be employed, some good can be accomplished by using arsenate of lead, 10 to 14 pounds in 100 gallons of water.

REFERENCES

Fernald & Kirkland, The Brown-tail Moth. 1903.
U. S. Farmers' Bull. 264. 1906.
U. S. Bur. Ent. Bull. 87. 1910.
U. S Farmers' Bull. 564. 1914.

Climbing Cutworms

The sleek, plump, dull-colored, obscurely marked caterpillars (Fig. 151), ranging from one to nearly two inches in length

FIG. 151. — One of the climbing cutworms, *Porosagrotis vetusta* ($\times 1\frac{1}{3}$).

and commonly known as cutworms, attack nearly all kinds of field and garden crops, and some of them often climb fruit trees, bushes and grapevines at night to eat the opening buds. More than a dozen different kinds of these climbing cutworms have been caught at their destructive work and identified as follows:

The yellow-headed cutworm (*Hadena arctica* Boisduval).
The white-spotted cutworm (*Homohadena badistriga* Grote).

The variegated cutworm (*Peridroma margaritosa saucia* Hübner).
The dark-sided cutworm (*Paragrotis messoria* Harris).
The white cutworm (*Paragrotis scandens* Riley).
The well-marked cutworm (*Noctua clandestina* Harris).
The black-lined cutworm (*Noctua fennica* Tauscher).
The mottled-gray cutworm (*Rhynchagrotis alternata* Grote).
The red cutworm (*Rhynchagrotis placida* Grote).
The speckled cutworm (*Mamestra subjuncta* Grote and Robinson).
The dingy cutworm (*Schizura ipomœœ* Doubleday).
The spotted-legged cutworm (*Porosagrotis vetusta* Walker).
A species of Prodenia and *Noctua baja* Fabricius.

Cutworms develop from eggs laid by night-flying Noctuid moths that are frequently attracted to lights in large numbers. Light, loose soils are most often infested by these caterpillars, and where there is a scarcity of low-growing vegetation they will climb almost any plant, even to the tops of high trees. As peaches are often grown in such light, sandy soils, both young and old trees have suffered severely from cutworms in various parts of the United States. The buds and leaves of grapevines are also favorite delicacies for them. Young apple, pear and cherry trees, or blackberry, raspberry and currant bushes, or young shade trees and shrubs grown in such soils are also often attacked. The half or two-thirds grown cutworms, hungry after a long winter's fast in the ground, emerge early in the spring as soon as the buds begin to open. Like thieves in the night, they crawl up the trees, vines or bushes and from about 8 P.M. until nearly morning continue their destructive work of eating the buds. In some instances the culprits have been first discovered on still nights by hearing the noise made by the clicking of the hundreds of tiny, hungry jaws as they devoured the buds. Fifty cutworms have been found at one time on a tree set the previous year; from 500 to 800 have been counted going up the trunk of a 12–year old apple tree in a single night; and 1500 have been taken from such trees during the 2 or 3 weeks they work in spring. Young

trees and grapevines are often stripped of buds and killed in a
single night, and the cutworms sometimes gnaw off the tender
bark of the twigs,
or may even girdle
the trunk if pre-
vented from ascend-
ing by some barrier.
Towards morning
they drop to the
ground, burrow in
an inch or more, and
remain during the
day. Usually in 2

Fig. 152. — Moth of *Porosagrotis vetusta* (× 1⅔).

or 3 weeks, or by the time the trees are in leaf and blossom,
the cutworms become full-grown, cease feeding and soon transform to the parent moths (Fig. 152).

Remedial measures.

Orchards or vineyards on the heavier soils are rarely troubled
by climbing cutworms. On the light, sandy soils usually preferred by these pests, keep the ground entirely free from all
grass and weeds for 2 or 3 months after July 15, so as to starve
out the recently hatched caterpillars. If some cover crop,
like rye, oats, clover, rape or cow-peas, could be sown late in
fall between the rows of trees, vines or bushes, and plowed under
after these fruits were in leaf, it would furnish the cutworms
something besides fruit-buds to eat and thus prevent much of
their destructive work.

A collar of cotton batting or wool properly put on the trunks
of trees or grapevines makes almost a perfect and a very cheap
barrier to the ascent of the cutworms. Unroll the batting into
thin sheets and cut into strips 4 or 5 inches wide. Wrap these
around the trunks, letting the ends overlap an inch or more,
then tie with common white twine *at the bottom* and carefully
roll the top of the band down over the bottom edge, thus form-

ing an inverted cotton batting funnel around the trunk. These barriers are not easily matted down by rains and a few hours of sunshine makes them as fluffy and effective as ever. In very rainy seasons use wool, which can be put away and used for several seasons.

Entice them away or prevent their getting to the buds by the above methods, then proceed to kill the culprits. Go out with a lantern at night, pick off the few that are able to reach the buds and collect or crush those trying to get up the trunks. As most of them can be found during the day just beneath the surface of the soil within a radius of a foot or two around the base of the trees or vines, it is an easy matter to dig them out. The fat, sleek caterpillars will make dainty morsels for a flock of chickens or other fowls. Many of these climbing cutworms can also be poisoned with a mixture of bran (25 pounds) and Paris green or white arsenic (2 pounds) and water enough to make a soft mash. Cheap sugar or molasses may be added to prevent it drying out so quickly, but apparently does not add to its attractiveness. First put on the cotton batting barriers, then drop a few large spoonfuls of this poisoned mash around the base of the trees or vines at nightfall. Many of the hungry cutworms will eat this deadly mash, either before they try to go up the trees or vines, or discouraged by the barrier, return and feast on it. Sometimes 90 per cent of them can be poisoned in this way. Of course, all poultry and other domestic animals must be kept away from the places where this poison bait is being used.

REFERENCE

Cornell Agr. Exp. Sta. Bull. 104. 1895.

CHAPTER IV

APPLE INSECTS — APHIS, SCALES, AND OTHERS

Plant-lice or Aphids

THREE different kinds of these small, soft-bodied sucking insects may attack the opening buds, the foliage and sometimes the young fruits on apple trees (Fig. 153). Often two species of these aphids work on the same tree in the spring. Nursery stock is frequently seri-

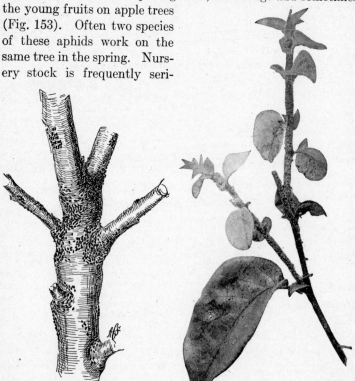

FIG. 153. — Aphid eggs on apple twig.   FIG. 154. — Apple leaf aphis on quince.

142

ously injured, the leaves being badly curled and the growth of
the tree checked. The aphids secrete a sweet liquid known as
honey-dew in which a black fungus de-
velops and gives the infested foliage and
twigs a sooty appearance. In large, bearing apple orchards, these plant-lice sometimes appear in incredible numbers and
not only curl and kill some of the terminal leaves but attack the young fruits,
preventing their full development and
giving them a knotty, stunted appearance (Figs. 157, 158 and 159). The three
species of these aphids now common on
apple trees throughout the United States
are *Aphis pomi, Aphis sorbi* and *Siphocoryne avenæ.*

FIG. 155. — Newly
hatched aphids clustering
on an opening apple bud.

The interesting and somewhat complicated life histories of these plant-lice have
not been fully worked out. The last two
species breed on the apple trees for only a few generations in
the spring and early summer, then migrate to other food-plants,
but return to the trees in the fall, where the hibernating eggs
are laid. The first species remains on the trees throughout
the year. The wonderful rapidity with which these aphids
multiply parthenogenetically enables even those that work on
the trees only a part of the season to develop in sufficient numbers to do serious injury to the foliage or fruit before they migrate to their summer food-plants.

Natural enemies of the apple aphids.

Cold, heavy rains, both in early spring and late autumn, remove and doubtless kill many of the aphids, thus often preventing serious infestation in orchards. Fungous diseases also
destroy large numbers under favorable weather conditions.
Several species of ladybird beetles, aphis-lions and maggots of

Syrphus flies often reap a rich harvest of apple plant-lice. These predaceous enemies, aided by several minute parasitic flies, are very effective aids in reducing these prolific little pests to much less injurious numbers. If it were not for these adverse weather conditions, fungous troubles, and insect enemies, apple

Fig. 156. — Dipping the tips of nursery trees in soap solution for the control of plant-lice.

trees, especially large, bearing trees, would be much more frequently overrun and severely injured by plant-lice.

Remedial treatments.

As all three of these common apple plant-lice pass the winter as shiny black eggs, quite readily seen when numerous, and laid mostly on the twigs, many of them could be removed by judicious pruning, especially on young trees. Nursery stock is frequently badly infested, and the dormant trees may be heavily stocked with the eggs. It is sometimes practicable to crush the eggs on a few young trees with the fingers or a thin wooden paddle. The eggs are very resistant to the strongest contact insecticides, like oils and soaps. Experiments indicate that spraying to kill the eggs is of doubtful utility, usually enough eggs hatching to abundantly stock the trees with aphids. Thorough fumigation with hydrocyanic acid gas is said to kill many of the eggs.

After the aphids hatch in the spring, they are readily killed *when hit* with " Black Leaf 40 " tobacco extract, $\frac{3}{4}$ of a pint in 100 gallons of water, adding 3 pounds of soap

FIG. 157. — Plant-lice clustering on a young apple.

to each 100 gallons to make the liquid stick and spread better. One thorough application when the aphids are thick on the opening buds will usually control the more common apple bud aphis, *S. avenæ*, which as a rule does not curl the leaves as much as the other species. The apple leaf aphis and rosy apple aphis, however, breed on the trees longer and often curl the leaves, so that it is almost impossible to hit a majority of them with the spray. Nurserymen often dip the infested branches into the insecticide (Fig. 156), thus doing more effective work than can be done with a spray. Several dippings or sprayings

L

are often necessary to thoroughly control these prolific little pests. Usually the weather conditions in the spring, aided by the insect enemies, prevent serious injury by the aphids on large, bearing trees. When they do infest such trees in destructive numbers and swarm on to the young fruits, the tree should be

Fig. 158. — Plant-lice clustering on young fruit.

promptly and most thoroughly sprayed with one of the insecticides mentioned above. Such sprays applied in October would kill many of the sexual forms before the eggs are laid. Nurserymen especially could thus strike a very effective blow at these plant-lice and often prevent much of their destructive work the next season.

THE APPLE LEAF-APHIS

Aphis pomi De Geer (*Aphis mali* Fabricius)

This old and common European species was not definitely recognized in America until 1897, but it is now widely distributed throughout the United States. It infests the apple, pear, quince, and hawthorn. From the small, shiny black oval eggs (Fig. 154) laid mostly on the bark around the buds in the fall by the wingless female aphids, there hatches in the spring, about the time the buds begin to open, the so-called stem-mothers. These are wingless, somewhat pear-shaped, bright green in color, and give birth to a generation of green viviparous aphids, about three fourths of which develop into winged females, the remainder being wingless with long, black cornicles.

FIG. 159.—Mature apples, dwarfed and misshapen, as a result of aphis injury when small.

The winged forms (Fig. 161) spread the species to other parts of the same tree or to other apple trees. About half of the next generation, and some of the later generations, may develop wings and migrate, but the winged forms give birth to wingless viviparous females only (Fig. 160). This species lives on the apple tree all the year, breeding continuously during the summer. Most of these wingless, viviparous females are light green in color, but in the spring some may have bright yellow bodies. In October a generation of true males and females, which are wingless, appear, mate and continue to lay their shiny black eggs on the bark for a

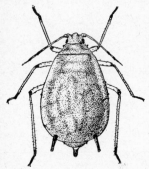

month or more. The light green, oviparous females have peculiar sensory pits on their hind tibiæ and are about two thirds as large as the parthenogenetic wingless summer forms. The yellowish-brown males, with blackish antennæ longer than the body, are one third smaller and much less numerous than the females.

This species usually appears somewhat later in the spring than the more common apple bud aphis, and it is thus not so numerous on the

Fig. 160. — Apple leaf-aphis, apterous viviparous female.

buds, waiting until the leaves are unfolded. As it breeds on the trees during the whole season it may be more injurious than the other two species. Its work resembles that of

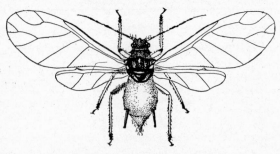

Fig. 161. — Apple leaf-aphis, winged viviparous female, third generation.

the rosy apple aphis, the leaves often being curled very badly. In 1907 it was so numerous in many large, bearing orchards in New York that it swarmed on to the young fruits in June, checked their growth, prevented the normal June drop, and gave them a knotty appearance, thus ruining them for market. Apples dwarfed by aphis injury usually present a characteristic puckered condition at the blossom end (Fig. 159). The rosy apple aphis also helped in this destructive and unusual outbreak.

REFERENCES

N. J. Agr. Exp. Sta. Bull. 143. 1900.
Del. Agr. Exp. Sta. 13th Rept., pp. 130–136. 1902.
U. S. Bur. Ent. Circ. 81. 1907.
Col. Agr. Exp. Sta. Bull. 133, pp. 23–28. 1908.

THE ROSY APPLE APHIS

Aphis sorbi Kaltenbach (*Aphis malifoliæ* Fitch)

This species is now widespread and common throughout the United States and Canada. It is probably an old European species which was introduced into America more than half a century ago. Apple is its favorite food-plant, but pear, white thorn and three species of *Sorbus* are sometimes infested.

This rosy aphis often occurs on the same trees and in the midst of colonies of the next species, *S. avenæ*, but it is not so restless and active. The stem-mothers hatched from the hibernated shiny, black, oval eggs in early spring, are globose in shape and of a dark purplish-brown color mottled with black. They are thinly covered with a whitish pulverulence and have blackish antennæ, cornicles and legs. The progeny of these stem-mothers are wingless parthenogenetic females, usually of a pinkish color, but sometimes varying to a light brown, slaty gray or greenish black with the body covered with a whitish coating. This mealy appearance and its pinkish color will usually readily distinguish the wingless forms of this rosy aphis from those of the other two species. The tips of the antennæ and cornicles are black. Another brood of these wingless viviparous females is developed on the leaves in early June, but the progeny of these, or the third generation from the stem-mothers, develop into parthenogenetic winged forms (Fig. 162) that migrate from the apple trees to some unknown food-plant during the latter part of June in New York. In 1893 we tried to colonize these spring migrants on various grasses, but

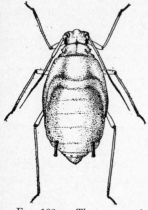

failed, and did not see the species again until September, when shiny black winged viviparous females or migrants appeared on the apple leaves. These were darker than the migrants which left the trees in the spring and differed slightly in other details. Early in October a progeny of globose, light yellow or brownish colored wingless oviparous females, with many sensory pits on the hind tibiæ, began to be born from the winged return migrants. Soon winged males, which resembled the somewhat larger return migrants, came from unknown sources and mated with the wingless females, which began laying their shiny black eggs on the twigs, and often on the trunk also.

Fig. 162. — The rosy apple aphis, a parthenogenetic female of the third generation, with wing pads.

Although working on the apple tree only about two months in the spring, this rosy apple aphis is capable of doing much injury. It often curls the leaves as badly as the preceding species, *A. pomi*, and in 1903 it helped this species in its very destructive work on the young fruits. In 1907 another of these unusual devastations on the fruit by plant-lice occurred in June in New York apple orchards, and this time the principal depredator was this rosy apple aphis.

References

Del. Agr. Exp. Sta. 13th Rept., pp. 149–156. 1902.
U. S. Bur. Ent. Circ. 81. 1907.

The Apple Bud-aphis

Siphocoryne avenæ Fabricius

This green aphis is the one which most commonly infests the opening apple buds in the United States and Canada. It is an old European species, which has a very wide range of food-plants, including apple, pear, hawthorn, quince, plum, at least seven other trees, five weeds or herbs, wheat, rye, oats and nine different grasses. It is most injurious to the apple, often nearly covering the opening leaf-buds and blossoms, and it sometimes injures young wheat in the fall.

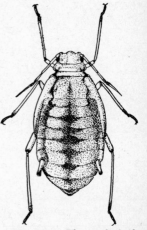

The yellowish-green, wingless, viviparous stem-mothers hatch from the shiny, black, oval winter eggs as soon as the apple buds begin to open, and most of their progeny develop into winged parthenogenetic, blackish females (Fig. 163) which leave the apple in May. A few may remain on the trees through two or three more generations, or until July. In 1893 we found that

FIG. 163. — The apple bud-aphis, a parthenogenetic female with wing pads.

these spring migrants could be readily colonized on June and meadow grasses (species of *Poa*), and we succeeded in following the insect through thirteen parthenogenetic generations on these grasses from late in May until November. Then some developed into the winged return migrants, but others continued to breed and finally lived through the winter on these grasses and on wheat kept under outdoor conditions. Curiously enough, all the generations grown on the grasses were alternately wingless and winged viviparous females. They were much smaller and darker colored than those on the apple, and have been described as a different species, *Aphis annuæ* Oestlund. On the grasses they lived on the blossom heads, but mostly on the stems, and some of them at the base of plants. It has been suggested that the species is biennial, the progeny of the spring migrants from the apple living on grasses and grains until the autumn of the second year before going back to the apple.

Late in September we found many winged viviparous females returning to the apple trees. These return migrants were very similar to the spring migrating form, and soon gave birth to oviparous females, which were wingless and of a yellowish-green or dark green color. About the time these females matured, or three or four weeks after the return migrant females came from the grasses, there came to the trees the more slender, light greenish-brown colored winged males. These males actively seek the wingless females, and we have seen them mate with the females of the rosy apple aphis also. Egg-laying began late in October and continued until December, the shiny black eggs being deposited on the bark all over the tree, but mostly on the twigs near the buds. The sexual forms may also appear on the pear, quince, hawthorn or plum trees in autumn, and eggs be deposited thereon.

All of the winged forms of this apple bud aphis, including the summer broods on grasses, can be distinguished from those of the other two species by the fact that the terminal fork of the second vein behind the stigma is shorter and nearer the margin of the wing. Usually the brownish coalescing spots around the clavate cornicles are more distinct on all the forms of this species. It leaves the tree for its summer food-plants so soon in the spring that it does not curl the leaves as much as the other two species.

References

Del. Agr. Exp. Sta. 13th Rept., pp. 137–149.   1902.
U. S. Bur. Ent. Circ. 81.   1907.

The Woolly Aphis of the Apple

Schizoneura lanigera Hausmann

Practically wherever the apple is grown in any part of the world, there may often be found during the summer on the trunk, branches and water-sprouts above ground and on the roots also bluish-white, cottony patches (Fig. 164) consisting

Fig. 164. — The woolly aphis, a cluster of lice on an apple twig showing the white woolly covering.

of many small, reddish-brown plant-lice or aphids scarcely one tenth of an inch in length. Above ground the bodies of the aphids are nearly covered by a woolly mass of long, waxy fibers that are much shorter on the root-inhabiting aphids, and gives them a whitish mealy appearance. Although for many years considered as distinct species and now often discussed as different forms, the aphids living underground on the roots and those on the branches or trunk are absolutely identical structurally and one can readily colonize the branch-inhabiting aphids on the roots and *vice versa;* furthermore, the aphids may often be seen during the growing season wandering from roots to branches or going down the trunk on to the roots, and in either case soon establishing themselves in their new location.

This woolly aphis has ranked as a serious apple pest for more than a hundred years both in Europe and America. In spite of much discussion and controversy, it is not definitely known which of these countries is its native home; it is found upon native apple and thorn trees in both countries. In America the insect is commonly known as the woolly aphis, but in

FIG. 165. — The woolly aphis, a cluster of lice with the woolly covering removed.

England it is the "American blight" and in Germany it is called the "blood-louse" from the red color of the crushed bodies of the aphids.

The woolly aphis sucks its food from the tissues of the bark and often causes an abnormal growth or swelling where it works. Above ground colonies of the aphids often develop about the leaf axils on sprouts or new growths and particularly at abrasions (Fig. 165) on the bark or where a branch has been cut off. The aphids often prevent the injured bark from healing normally, and as considerable enlargements of the surrounding tissues result, infested branches often present a swollen and scarred appearance. A favorite location for their work is on the crown of the tree just above the roots. Underground the aphids cause conspicuous, rounded, nodular swellings or galls to develop on the roots, finally resulting in their decay. Usually most of the injury resulting from the work of this pest is due to the greater numbers of the aphids infesting the roots, the more conspicuous but less numerous colonies above ground, rarely doing much damage, especially on larger trees. In some regions, and especially in Europe, the trees are often seriously injured by a majority of the aphids working above ground. In cases of severe infestation the woolly aphids swarm over the whole tree above and below ground, even attacking the foliage and fruit. The foliage on badly infested trees often presents a yellowish, sickly appearance, and the trees are easily uprooted, as many of the roots have decayed from the work of the pest. Apple trees of all ages and varieties are liable to attack, but usually young trees, especially nursery stock, suffer most. Some varieties, like the Northern Spy, are often more or less exempt from attack. Pear, quince and the mountain ash are also recorded as host-plants; it also passes a part of its life history on the elm, causing a characteristic curling of the leaves (Fig. 166). The insect may injuriously infest the roots of trees growing in various kinds of soils, and it works destructively over a wide range of latitude. Many thousands of nursery trees are annually either killed by the insect or rendered unsalable and destroyed in America. Infested nursery stock is largely responsible for its wide distribution. In most localities large, thrifty orchard trees are not seriously injured by this aphis, but sometimes under favorable conditions it breeds so rapidly that it ranks among the most destructive of the insect enemies of the apple.

The interesting and rather complicated life history of this woolly aphis is little understood by orchardists, and a few details are still lacking to make it complete. During the summer only the little wingless, agamic female aphids occur on the apple trees. A dozen generations of these may be developed during the summer, each mother aphid bringing forth living young, sometimes at the rate of two to twenty a day for two or more weeks. The baby aphids or nymphs are usually born enwrapped in a thin pellicle, which is soon cast off. The little creature begins to suck its food through a beak longer than its body, and its waxy coating is secreted in a few hours. As these little nymphs feed and grow their skin is shed four times, a new waxy coating being secreted each time, and they may become full-grown in from eight to twenty days. Many of these wingless, agamic nymphs persist on the roots, and some of them even on the tree above ground, all the year through even in New York state and other cold northern latitudes. These aphids mostly, if not wholly, cease breeding, however, even in southern localities during the winter months. During the autumn months, sometimes beginning in August, there is developed both above and below ground many minute, winged, greenish-brown-bodied, agamic female aphids with the body more or less covered with the woolly secretion. These winged forms may fly or be blown to near-by elm trees. They are destined to play an interesting and important rôle in the perpetuation of their kind. In a few days these winged, agamic migrating forms give birth to from six to twelve young, about half males and half females. Both sexes are wingless and do not grow after being born, having no mouth parts with which to take food. The reddish-yellow females are about one-twentieth of an inch in length and twice as large as the slenderer, olive-yellow males. A few days after mating the female lays a single long, dark, cinnamon-colored oval egg nearly as large as her body in a crevice of the elm bark. Some of these eggs have been found in crevices of the apple bark where there had been colonies of the lice during the summer; others record them as laid on the bark on the crown of the tree near the roots, but as a rule they are laid on the elm. These winter eggs hatch in early spring and the stem-mothers, as the first brood of lice are called, are found on the opening elm leaf buds. They are wingless, and feed on the under surface of the leaves and are soon surrounded by a numerous family of young aphids. The presence of the lice cause the elm leaves to swell and curl, as shown in Figure 166. The next generation is also wingless, but with the third brood winged forms appear and continue abundant throughout the summer. Some of these fly back to the apple and there establish colonies on the branches, others probably found colonies on the tender elm branches, and some of those appearing early in the season may migrate to other elm leaves. When living on the elm the woolly aphis has been known as the woolly elm leaf aphid (*Schizoneura americana* Riley).

FIG. 166. — Elm leaves curled by the woolly aphis.

Enemies.

The woolly aphis has its natural enemies, which help to keep it in check. Spiders often spin their webs over a colony of the aphids and then live at their ease. Many of the aphids are parasitized by the minute chalcis fly, *Aphelinus mali*, and the larvæ of lace-wing flies and a syrphus fly, *Pipiza radicum*, often work destruction in the woolly clusters. The larvæ and adults of several ladybird beetles, particularly the small, brown *Scymnus ceryicaulis*, and the nine-spotted *Coccinella 9-notata*, are also active enemies of the woolly aphis, but the combined efforts of all these foes do not often sufficiently control it, so as to make remedial treatments unnecessary, especially on young trees.

Remedial measures.

Above ground the woolly aphis can be readily controlled by thoroughly drenching the bark, and particularly the woolly colonies of aphids in summer, with a forceful spray of 15 per cent kerosene emulsion. Two applications may be necessary in extreme cases. Soap solutions and tobacco decoctions are not so effective as the emulsion. Drench the lower portion of the trunk and let the spray run down on to the crown and roots. It is often practicable to destroy most of the aphids above ground by simply painting the woolly patches with the emulsion or pure kerosene, crude oil, or a miscible oil (1 to 10 parts water).

It is much more difficult to reach and kill the aphids working underground on the roots. Nurserymen should destroy all badly infested stock where the roots show many of the characteristic galls. Orchardists should never accept and plant such trees. Where infestation by this pest is suspected, all stock should be properly fumigated with hydrocyanic-acid-gas, or the whole trees or the roots only may be dipped in the 15 per cent kerosene emulsion mentioned above, or in a nicotine solution. For dipping, the roots should be freed from lumps of dirt, then held in the liquid a minute or two and spread out to dry before being piled in heaps. Hot water (130° to 150° F.) is said to be an effective dip also, but do not use the lime-sulfur wash, as it often injures or kills the trees. If the emulsion is used, it must be thoroughly emulsified, as any free oil might seriously injure the roots. Plant the trees free from the pest, keep them growing thriftily for a few years, and the woolly aphis will be much less liable to seriously infest the orchard.

A narrow band of some sticky material around the trunks of infested trees will capture many of the aphids often seen wandering up and down the trunks, and thus aid materially in preventing the infestation of the roots by those working above ground.

Experiments in Missouri in 1896 seemed to show that the aphids could be killed and their ravages largely prevented by a liberal use of tobacco dust. It was applied on a 10-year-old orchard by removing the earth to a depth of 4 inches and for a distance of 2 feet around the trunk. Five or six pounds of the dust were evenly scattered over this area and the dirt replaced. Nursery trees were treated by putting the dust in trenches alongside the rows. Further detailed experiments in Georgia ten years later, however, gave very unsatisfactory results in orchards and nurseries with tobacco used in various forms in the excavated areas or trenches. Some trees received over 12 pounds of the dust in four months, two applications being made. Whale-oil soap was also used in these experiments even at the rate of 2 pounds to a gallon without success; and it was found that carbon bisulphide injected into the ground would kill the aphids only over a limited area near the application hole, and it could not be used in sufficient quantities to kill all the aphids without killing or injuring the trees. These Georgia experiments demonstrated the value of an application of 15 per cent kerosene emulsion (the stock emulsion formula diluted with about 10 gallons of water). The soil was removed to a depth of about 3 inches over an area from $1\frac{1}{2}$ to 4 feet around the trees, depending on the size of the trees, and after the application the soil was replaced. Three gallons of the 15 per cent emulsion on the smaller area and 6 gallons over the larger circles served to saturate the soil for 2 to 4 inches, and it gradually permeated the soil a foot or more, where a perceptible odor remained for many weeks. All the aphids the emulsion reached were killed and the kerosene odor acted as a repellent for a long time. The cost per tree varied from 4 to 8 cents. A 10 per cent emulsion (stock formula diluted with 17 gallons of water) was very effective on nursery trees when poured in shallow trenches made close to the trees along each side of the rows. This kerosene emulsion treatment should be made during early summer and not later than the last of July, as it may injure dormant trees or those that have made most of their growth for the season.

As there is danger that nursery trees may become infested by migrants from elm, it is not advisable to allow these trees to grow in the vicinity of the nursery or to grow elm stock with apple stock in the same nursery.

REFERENCES

Mo. Agr. Exp. Sta. Bull. 35. 1896.
U. S. Bur. Ent. Circular 20. 1897.
U. S. Bur. Ent. Bull. 18, pp. 78–81. 1898.
Ga. State Bd. Ent. Bull. 23. 1907.
Col. Agr. Exp. Sta. Bull. 133, pp. 5–23. 1908.
Me. Agr. Exp. Sta. Bull. 203. 1912.
U. S. Bur. Ent. Circular 158. 1912.
Me. Agr. Exp. Sta. Bull. 217. 1913.
Me. Agr. Exp. Sta. Bull. 220. 1913.

THE BUFFALO TREE-HOPPER

Ceresa bubalus Fabricius

Young fruit trees in the upper Mississippi Valley and east-ward through the United States and Canada to Nova Scotia are often seriously injured by this curious little grass-green, triangular-shaped, active bug, about ⅜ of an inch long, with large, horn-like projections of the anterior angles of the thorax which give it a fancied resemblance to a male buffalo, hence its common name (Fig. 167). The injury is all done by the fe-male bugs in laying their eggs from July till October in the bark on the upper sides of the smaller branches. Two slightly curved slits about $\frac{3}{16}$ of an inch long are made near together in such a way that the bark between the incisions is cut loose. From 6 to 12 long, cylindrical, whitish eggs are stuck into each slit and do not hatch until the following May or June. These peculiar egg-scars rarely heal and gradually enlarge, giving the branches a very rough and scabby appearance (Fig. 167 *a*). Two- or three-year-old apple and pear trees usually suffer most, but cherry, prune and quince trees and various forest trees often show many of the egg-scars. Hundreds of the incisions are often made in a square inch or two of the bark, growth is checked and such extensive scarification often ruins the trees. In some localities the insect is considered the most destructive insect enemy of young fruit trees.

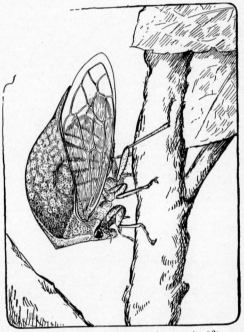

The strongly spined young bugs or nymphs which hatch from the eggs live mostly on grasses or weeds near the scarred trees. Thus only uncultivated orchards or those bordered by low vegetation are seriously injured by this buffalo tree-hopper. Two minute parasites destroy many of the

Fig. 167. — The buffalo tree-hopper ($\times 5\frac{2}{3}$).

eggs. Thorough cultivation and the burning over of weedy borders in June will starve out and largely prevent injury from this sucking bug. It cannot be reached satisfactorily with any spray, but many of the eggs can be destroyed by judicious pruning out of the freshly scarred branches in autumn or winter.

Stictocephala inermis Fabricius, a species of tree-hopper similar to the last, also scars the branches with its characteristic

M

egg punctures. It causes little injury, however, because the inner bark continues alive and there is no dead area between the slits.

Ceresa taurina Fitch and *C. borealis* Fairmaire, two forms closely related to the buffalo tree-hopper, deposit their eggs in

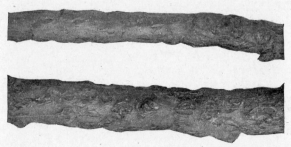

Fig. 167 *a*. — Apple twigs showing egg-scars of the buffalo tree-hopper.

the buds, within the outer bud-scales. They cause no appreciable injury.

REFERENCE

N. Y. (Geneva) Agr. Exp. Sta. Tech. Bull. 17. 1910.

THE SAN JOSÉ SCALE

Aspidiotus perniciosus Comstock

The San José scale has attained greater notoriety, has been the cause of more legislation, both foreign and interstate, and has demonstrated its capabilities of doing more injury to the fruit interests of the United States and Canada than any other insect. The ease with which it is widely distributed on nursery stock, the practical impossibility of exterminating it in a locality, its enormous fecundity enabling it to often overspread the bark, leaves and fruit of trees in a very few years, and the fact that it attacks practically all deciduous fruit and ornamental plants, makes it of the greatest economic importance. No other scale-insect has ever equaled it in capacity for injury to plants.

China is believed to be the native home of this pest, and more appropriate common names for it are the Chinese scale, or the pernicious scale, from the very pat name given it by Professor Comstock when he described it in 1880. It first became established in America at San José, Cal., about 1870, and derived its name therefrom. Previous to its introduction into Eastern nurseries in 1886 or 1887, the scale had gradually spread over most of the states west of the Rocky Mountains. It was not until 1893 that it was discovered in Virginia in the East, but it was soon found to have been already widely spread from these nurseries through thirteen states from New York to Florida. So rapidly has it been spread that important orchard sections in nearly every state and territory, and in Canada and British Columbia, are infested, and it is only a question of time when it will extend over practically all the fruit-growing areas of North America within its climatic range. It occurs also in Hawaii, Chili, Japan and Australia, but stringent legislation has thus far prevented its becoming established in Europe.

The San José scale attacks all parts of fruit trees, including the trunk, branches, leaves and fruit, and usually causes reddish discolorations of the bark or skin of fruit (Fig. 169). Clusters of the scales often occur around the stem and blossom end of the fruit, rendering it unsalable, and sometimes giving a pitted appearance. In bad infestations the scales are crowded together and present a grayish, roughened, scurfy deposit on the bark. If scraped, a yellowish liquid results from the mashing of the soft yellow insects beneath the scales. The fruits commonly infested are apple, pear, quince, peach, plum, prune, apricot, nectarine, sweet cherry, currant and gooseberry. Lemons and oranges, except the trifoliate varieties, many shrubs, forest trees and evergreens are practically exempt from attack.

The fears that shade trees and forests would be ravaged

and become permanent breeding grounds have not been realized, as the pest confines its depredations mostly to fruit trees and ornamental shrubs.

The scale is a waxy secretion covering the soft, yellow,

sac-like body of the insect beneath. The largest scales cover the full-grown females and are nearly circular, gray, about the size of the head of an ordinary pin ($\frac{1}{16}$ of an inch in diameter) with a central dark nipple surrounded by a yellowish ring (Fig. 167 *b*). The smaller scales are nearly black with a central gray dot surrounded by a black depressed ring bordered by a grayish ring. The San José scale can often be readily

Fig. 167 *b*.—Full-grown female San José scale (× 10).

distinguished from the closely related species, Putnam's scale, European fruit-scale and cherry scale, even with a hand-lens by these peculiarities of the young scales. In the other species the nipple is usually one side of the center and orange or yellow in color, and the scales are not so black or lack the depressed ring of the nipple. The elongate-oval

male San José scales, only about half as long as the diameter of a mature female scale, are dark gray with the circular raised ex-uvial portion near one end and usually darker but sometimes yellowish (Fig. 168). The male scales are sometimes more numerous than

Fig. 168.— Two male San José scales. Enlarged.

the females during the early part of the breeding season.

In late autumn all stages of the San José scale, from those just born to the fully developed insects, are to be found on the trees, but practically only the small black scales covering the

half-grown insects hibernate, all the other stages being killed
by winter conditions. In New York the winged males emerge
in May, and the females mature and begin giving birth to living
young during the latter part of June. The young develop
inside the body of the mother in thin membranous, sac-like
eggs and most of them burst through the sac and are born alive,
but some of these eggs may be laid before the young hatch,
so while the insect is usually ovoviviparous it may be partially
oviparous. A single mother is capable of giving birth over a
period of six weeks to nearly 600 young but doubtless does not
average more than 100 to 200 ; many of these are males, and some
soon die. The tiny yellow, six-legged young crawl from under
the mother scale and often spend about a day in finding a suitable place to settle down and insert their long, thread-like mouth
parts with which they suck their food from the interior tissues
of the plant. In a few hours the body becomes covered with
a mass of white cottony and waxy fibers which in 2 or 3 days
mat into a pale grayish scale that gradually becomes larger and
darker until in about two weeks the first molt of the insect
occurs. Up to this point the males and females and their scales
have been indistinguishable, but after this molt they both lose
their legs and antennæ and the females their eyes also. The
males have large, purple eyes and undergo two more molts,
gradually developing into delicate, orange-colored, two-
winged, fly-like insects in from 3 to 4 weeks. The yellow
female insects, with their thread-like, sucking mouth parts two
or three times as long as the body, remain circular, flattened
and sac-like in form, molt a second time in from 3 to 5 weeks,
and in a few days mate with the males. In molting the old
skins split around the edge of the body, the upper half adhering
to the scale beneath the central nipple and the lower half forming sort of a ventral scale next to the bark ; the second and
third cast skins of the male are pushed out from beneath the
scale.

At Washington, the females may attain their full growth in 30 days from birth, but it requires about 50 days in the fall in New York. As the females bear living young over so long a period, the broods overlap and it is difficult to trace the number of generations, but there are apparently three broods annually in the latitude of New York and four broods at least south of Washington. Breeding begins in the North in June and a month or more earlier in the South. The progeny of a single San José scale giving birth to only 100 female young in the spring could, but doubtless never does, amount to the enormous total of over 100,000,000 females by fall if there were four generations annually. At this fearful rate of multiplication, unequaled by any other injurious scale-insect, it is no wonder that infested plants rapidly succumb to the drain of so many thousands of tiny pumps sucking out their life.

The widespread distribution of the San José scale is due almost entirely to infested nursery stock. Rigid nursery inspection, compulsory fumigation, and interstate quarantine legislation doubtless help much, but fail to fully protect the fruit-grower, and the pest continues to reach both old and new localities on infested nursery stock bearing supposed "bills of health" in the form of inspection and fumigation certificates. Many have feared that new infestations might be brought about through infested fruits, especially apples and pears, which are distributed world-wide. Foreign nations enacted strict quarantine regulations against infested American fruit, even though it were dried. However, there are no authentic cases of infestation from scaly fruit, and while there is a bare possibility that it might occur, the chances are so small as to be practically ignored. The scale may be spread locally from tree to tree or to other orchards in several ways. As the newly born lice are active and often crawl about for a day before settling down, they may be able to crawl on to other trees, especially in nurseries where the branches interlace and touch. Strong winds may blow these crawling young to neighboring trees, and many of them are doubtless carried to other trees or orchards by other insects and by birds which often go from tree to tree. The crawling young have been found on the bodies of black lady-bird beetles, black ants, grasshoppers, *Chrysopa* adults, flies, and beetles, these insects thus furnishing ideal steeds or " flying machines " on which the scale may ride to new pastures.

While the San José scale is one of the greatest insect scourges that the fruit industry has ever encountered, it has taught some valuable lessons. Nurserymen are growing and shipping cleaner, healthier, better stock. Fruit growers are selecting their trees with greater care, and giving each tree individual attention in the orchard, an invaluable feature in orcharding. Many have been forced into spraying, which most progressive fruit growers find to be one of the best paying operations in orchards. The

FIG. 169. — Pear infested with San José scale.

scale is so small and so difficult to reach and kill, that the efforts to successfully combat it have resulted in better spray mixtures, machinery and methods not only for this scale, but for other insect and fungous enemies of orchards. Fruit growers in general are spraying more skillfully, more easily and more effectively, and many of them are satisfactorily controlling this tiny but terrible foe — the San José scale.

While the San José scale is beset by many natural enemies, its marvelous fecundity usually enables it to develop in injurious numbers in spite of them. The following nine species of minute Hymenoptera are true parasites of the scale in America: *Aphelinus fuscipennis* and *mytilaspidis*, *Aspidiotiphagus citrinus*, *Anaphes gracilis*, *Physcus varicornis*, *Prospaltella aurantii*, *P. perniciosi*, *Ablerus clisiocampœ* and *Rhopoidens citrinus*. Most of these parasites are widely distributed in the United States and other countries, and they are all general feeders on other species of the armored scales. Sometimes these parasites destroy enormous numbers of the scales, and they will always be very potent factors in Nature's efforts to help man in controlling this pest.

About a dozen ladybird beetles have been found eating the San José scale in America. The most important and useful of these are the twice-stabbed ladybird, *Chilocorus bivulnerus*, the tiny black *Microweisea misella*, another tiny, dark, wine-red colored species of the same genus, *M. suturalis*, and a Malachiid beetle, *Collops quadrimaculatus*. The most useful and interesting of these is the tiny black *misella*, which is widely distributed in the United States. The little beetles stand astride the full-grown female scales, push their heads under the margin of the scale and devour the soft, yellow insect beneath. The grubs of the beetle feed upon the smaller scales. The chief natural enemy which kept this scale in check in its native home in China was found to be a ladybird beetle, *Chilocorus similis*, which is almost identical in the beetle stage to our native American twice-stabbed ladybird, but differs in the reddish color of the grub and it also breeds much faster. This Asiatic ladybird beetle was introduced into the United States and readily attacked the scale, multiplied rapidly at Washington, and was sent into other localities both North and South. Lack of food and a native parasite destroyed the Washington colony and the insect failed to thrive in the North. It bred in great numbers for a time in Georgia, but man's spraying operations soon cut off its food supply and it was nearly exterminated.

Several fungous and other diseases sometimes attack the San

José scale with much effectiveness, especially in its southern range. Some of these can be transferred from tree to tree and it is in the range of possibilities, with more careful study, that artificial cultures can be made and distributed and yield results of practical value in the control of this scale. No harm can come from such introduction of natural enemies or diseases, and the time will doubtless eventually come, as it apparently has in some localities in California especially, when these enemies and diseases, together with man's vigorous warfare, will rob this insect of most of its present terrors, making it a much less dangerous orchard pest.

Remedial measures.

Its minute size rendering it difficult to detect unless very numerous, the ease and rapidity with which it may be distributed on nursery stock or cuttings, its marvelous fecundity enabling a few scales to soon re-infest a whole tree and the skill required to hit all the tiny scales with a spray, make the San José scale one of the most difficult insect pests to successfully control. Extermination is practically impossible except on limited areas where the infested plants can be destroyed root and branch, and then a new infestation may occur at any time if more plants are set. The destruction of infested trees is advisable in a very young orchard where only a few trees are involved, so as to put off as long as possible the general infestation, which will usually follow sooner or later. In older bearing orchards many fruit growers have succeeded in getting the pest under thorough control without the loss of a tree, but it means a big and continuous fight by a man determined to win.

Fruit growers should become familiar with the appearance of the scale from dead specimens readily obtained from infested localities. The pest can be combated more easily and effectively while it is in hibernation as half-grown scales on dormant trees. In starting new orchards get certified, fumigated stock from reliable sources, carefully examine each tree, and then thoroughly fumigate the stock again with hydrocyanic acid gas before setting. Fumigation with this gas, if properly and thoroughly done, is the most effective and practicable treatment for nursery stock. The dipping of such trees in the lime-sulfur wash or other sprays is not so effective and may injure the trees, especially if the roots are dipped.

After much experimentation with fumigation tents, whale-oil soap, undiluted kerosene and crude oil, and mechanically mixed oil-water sprays, these have been largely superseded by the cheaper, more effective and safer sprays of oil emulsions, miscible oils and the lime-sulfur wash, which must be brought in contact with each scale. Several applications of a 10 to 15 per cent kerosene emulsion spray during the summer has been safely used on apples to check the development of the pest, and a 25 per cent crude petroleum emulsion (formula, p. 486), makes an effective spray for use on old apple trees as the buds are swelling in the spring. The miscible oils should not be diluted more than 1 gallon of oil to 10 or 12 of water to get satisfactory results. The lime-sulfur wash should be used in preference to other sprays on peaches, for when applied in early spring it kills the scale and also acts as an effective fungicide against the destructive peach leaf-curl fungus. Badly infested trees should be sprayed twice, first in late autumn after the leaves drop and again in early spring before growth begins. Some are able to successfully control the pest with only one application in early spring. The cheapest and safest spray, the one which has withstood the severest tests of experimenters and orchardists, and has given the most uniformly successful results, is the lime-sulfur wash. The oil emulsions or miscible oils are non-corrosive, more agreeable to use, spread better, so that less material is necessary, and they penetrate more effectively the crevices of the bark or the fuzzy coated twigs of apple trees, but unless properly applied there is always more or less danger of injuring the trees. The market brands of miscible oils are simply poured into the required amount of water and quickly form, upon stirring slightly, a perfect and stable emulsion-like mixture ready for use. None of these sprays recommended for use on dormant trees can be safely used on trees in foliage. Where very large, old, rough-barked apple trees are infested, only the most thorough kind of spraying will conquer the pest. On such trees it is recommended to use the crude oil emulsion spray just as the buds are swelling in the spring.

The most effective work can be done with about 100 pounds pressure per square inch, using a fine spray through a nozzle of the cyclone type. The judicious pruning away of the tops and long sprawling branches of infested trees will often enable the orchardist to do more thorough work. Remember that the San José scale is not larger than a pin head; that the insect itself is well protected under the scale; and that it is therefore necessary to hit each tiny scale so thoroughly that the spray covers the insect. To do this requires powerful pumps, good nozzles and, most important of all, an experienced and determined man behind the gun who can shoot straight and thoroughly cover the bark of the tree with the spray from the surface of the ground to the tips of the smallest twigs.

REFERENCE

U. S. Bur. Ent. Bull. 62. 1906.

Nearly every Agricultural Experiment Station has published bulletins or circulars giving full directions for fighting the San José scale under local conditions.

THE OYSTER-SHELL SCALE

Lepidosaphes ulmi Linnæus (*Mytilaspis pomorum* Bouché)

This cosmopolitan insect is doubtless the commonest, most widespread and best known of the scale-insects infesting fruit-trees in America, where it has been injurious in the northeastern

United States for a century.   By 1850 it was abundant throughout the northern states east of the Mississippi.   Spreading

rather slowly, it has now reached most of the orchard sections in the far West, the South and all through Canada, but it is most injurious throughout its northern range from Nova Scotia southward to the latitude of Washington and westward to Montana.   In the North the scales often develop on the fruit itself, causing red spots similar to those produced by the San José scale (Fig. 174).

The oyster-shell scale is readily distinguished from all other scale-insects injuriously infesting deciduous fruit-trees in America by its peculiar shape and color, resembling a miniature elongate, curved oyster shell of a dark brownish bark-like color.   The convex scale covering the body of the female is about ⅛ of an inch long and consists of two minute cast skins at the smaller end and a large scaly portion gradually secreted from the body of the insect underneath.   The male scale is much smaller and rarely seen on fruit-trees; they are often abundant on ash.   Old lifeless scales often adhere to the bark for several years.

Fig. 170. — Apple branch badly infested with the oyster-shell scale.

If at any time from September to May the female scales formed during the preceding summer be overturned, they will be found to cover from 30 to 100 minute, white eggs and the much shriveled, dead body of the mother tucked away at the smaller end (Fig. 173). Thus hibernation in the egg stage lasts for 8 or 9 months, the time of hatching in the spring depending much on weather conditions. Hatching may begin as early as the middle of May in the North, but in 1907 it was a month later in New York. The mere specks of active six-legged, pale yellowish-white young (Fig. 171) that hatch from the eggs soon crawl out from under the scale and in a few hours settle down on the bark, insert their long, thread-like sucking tube, secrete

Fig. 171. — Apple branch infested with oyster-shell scale showing newly hatched lice.

a covering of cottony fibers, and the females never move from that spot (Fig. 172). The sexes are alike at birth, and after feeding a few days shed their skin, becoming grub-like creatures without legs or antennæ. Growth continues with no apparent difference between the sexes until it is necessary to molt again, when it is seen that a winged insect, the male, is being developed under some of the scales. This second cast skin of the female is added to the scale-covering; a few days later the fully developed, delicate, two-winged male insect without mouth parts emerges and seeks its mate. The yellowish females continue to increase in size, remain grub-like in form and secrete the large, brown portion of the scale, becoming full grown in August or early September in the North. Egg-laying soon begins, the body of the mother gradually shrinking into the smaller end of the scale, and the 30 to 100 eggs occupying most of the space beneath the scale. In New York, egg-laying sometimes begins early in August, but in 1907 it was delayed until October in some localities. There is but a single generation of the oyster-shell scale in the North, but in southern New Jersey and Pennsylvania and farther south there are two generations annually.

Fig. 172. — Old and recently set oyster-shell scales on willow.

This oyster-shell scale has a wide range of food-plants. It often nearly covers the bark of the larger branches (Fig. 170), and even the twigs of apple and pear trees, and is often equally as numerous on lilac bushes, willow, mountain ash and poplar trees. It may also attack quince, plum, raspberry, currant and fig among the fruits, and includes more than twenty-five shade trees and shrubs in its list of host-plants. It infests trees of all sizes and ages, often killing young trees and severely injuring large ones. Orchards that are kept in a thrifty growing condition and the trees not crowded rarely suffer serious injury from this scale, but we have seen the lower limbs especially, and sometimes the whole of large trees, killed by the insect where the trees were crowded and neglected. Usually the bark of the tree only is infested, but occasionally a few of the scales develop on the fruit even in the North, where there is but a single generation annually.

The oyster-shell scale is beset by many natural enemies. Some of the ladybird beetles, the twice-stabbed ladybird especially, devour many, and the eggs beneath the scales are preyed upon and often a large proportion of them, 50 to 75 per cent in some cases,

Fig. 173. — Oyster-shell scales turned over to show eggs.

eaten by a mite, *Hemisarcoptes coccisugus*, in France, and in America by the larvæ of at least five minute parasites, *Aphelinus mytilaspidis, abnormis,* and *fuscipennis, Anaphes gracilis,* and *Chiloneurus diaspidinarum.* These parasites

Fig. 174. — Apple infested with oyster-shell scales.

emerge through pin-like holes in the scales and often a majority of the scales on a tree show these holes. It usually requires two of the parasitic larvæ to destroy all the eggs under a scale, one larva often leaving from 2 to 20 eggs. A few birds, the brown creeper, black-capped chickadee and white-breasted nuthatch, are also reported as feeding on it. The combined efforts of all these natural enemies often prevent serious injury by the oyster-shell scale and occasionally nearly exterminate it in a locality.

Remedial treatments.

Orchardists should reject all nursery stock bearing many oyster-shell scales. Nearly three fourths of the life of this pest is spent as eggs well protected by the scale, and no thoroughly successful method has yet been found for destroying these eggs directly. Experiments indicate that fumigation with hydrocyanic acid gas, as practiced by nurserymen to kill the San José scale, often does not kill more than two thirds of the eggs of the oyster-shell scale. We have failed to kill many of the eggs even with undiluted kerosene.

The oyster-shell scale can be satisfactorily controlled, however, by thorough applications of lime-sulfur as recommended for the San José scale (p. 170). The lime-sulfur seems to loosen the scales from the bark so that the eggs are either blown away or fall a prey to their numerous enemies. This method is usually adopted by commercial growers and is the most practicable means of fighting this pest in large orchards.

Effective work can also be done against the oyster-shell scale in June, usually from the 1st to the 15th, when the young have recently hatched and are crawling about or have just begun to secrete a scale, by thorough spraying with a soap solution, 1 pound in 5 gallons of water, with " Black Leaf 40 " tobacco extract, $\frac{3}{4}$ of a pint in 100 gallons of water, adding 3 pounds of soap to each 50 gallons, or with kerosene emulsion made according to the standard formula and diluted with 6 parts of water. In the South, where two broods develop annually, the young can be killed with one of these sprays in August or September.

THE SCURFY SCALE

Chionaspis furfura Fitch

For more than half a century this native American insect has been one of the commonest and best known scales infesting pear and apple orchards in the northeastern portion of the United States, and its range now includes most of Canada and the United States; it has been introduced into England. It is often so abundant as to nearly cover the bark of apple, pear, currant, black raspberry, Japan quince and mountain ash, sometimes killing these plants, but usually it is not very destructive. Its different food-plants now number nearly twenty-five, including the peach and quince, besides the fruits just mentioned. Usually the bark only is infested by this scurfy scale, but rarely it gets on to the fruit and causes similar but larger reddish discolorations of the skin than the San José scale (Fig. 176).

Fig. 175. — Apple branch infested with the scurfy scale.

The rather flat, somewhat pear-shaped grayish-white female scales are about $\frac{1}{8}$ of an inch long and when numerous give the bark an ashy or scurfy appearance, whence the common name (Fig. 175). The scale consists of the two minute cast skins at the smaller end and a large, broad, thin, whitish portion secreted by the yellowish grub-like female insect beneath, and is, therefore, easily distinguished from the narrower, more convex, dark brown oyster-shell scale. The male scurfy scales are much smaller than the females, brilliantly white in color, with nearly parallel sides, three longitudinal ridges or keels, and a single conspicuous, yellowish-brown cast skin at the end

N

of the scales. The males are sometimes quite numerous on certain branches.

The details of the development and life history of this scale are very similar to those of the oyster-shell scale. There is evidently a single brood annually in its northern range, but two broods are recorded in Illinois, Ohio, Pennsylvania, Delaware and southward. In 1895 we reared in New York two generations of the scurfy scale and one generation only of the oyster-shell scale from eggs which hatched May 13 on trees in an insectary, and found corroborative evidence on trees in the field, but this was doubtless an exceptional season. The minute, purplish young which hatched May 13, molted first early in June, and the females the second time about a week later. By the middle of June the males had emerged and by July 9 many of the eggs were laid, and these hatched in about ten days. On August 22, a mature male and female in the same stage as those on the tree in the insectary, were seen in the field, and eggs were laid during the first three weeks in September. Owing to the purplish bodies of the young and half-grown scurfy scales and the thinness of the scale, they are not nearly so conspicuous on the bark as the oyster-shell species. The number of purplish-red eggs laid by the females varies from 10 to 85, averaging about

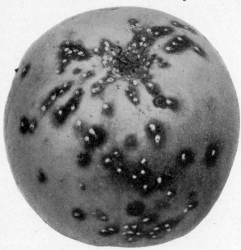

Fig. 176. — Apple infested with scurfy scale.

50, and these hibernate with the dead and shriveled bodies of the mothers under the scales.

The scurfy scale apparently has fewer enemies than the oyster-shell scale, as but a single parasite, *Ablerus clisiocampæ*, is recorded. Two ladybird beetles, the twice-stabbed ladybird and a species of *Hyperaspidius*, however, devour many of them.

Remedial treatments.

The remedial measures recommended for the oyster-shell scale will also control the scurfy scale.

PUTNAM'S SCALE

Aspidiotus ancylus Putnam

This native American scale insect is common and widely distributed, but is rarely injurious except upon plum trees and currant bushes, which are sometimes incrusted as badly as with the much more dangerous and destructive San José scale. Apple, pear, peach, nectarine, cherry and orange trees, besides several shade trees, are also among its food-plants. The mature, dark gray or blackish, nearly circular female scale (Fig. 177) can be distinguished from the European fruit-

FIG. 177. — Putnam's scale, enlarged. Redrawn from Joutel.

scale or the cherry scale by microscopic characters only, but with a hand-lens it is usually readily separated from the San José scale by its exposed orange exuvium situated at one side of the center. It passes the winter in a nearly full-grown condition. The males appear in April and the females are oviparous, depositing from 30 to 40 eggs under the scales in late spring or early summer. The crawling young lice hatch mostly in June and July, and there is but a single generation in a year in northern latitudes.

Parasitic enemies kill many of the insects under the scales. As the species spreads slowly, cut out and burn the worst infested branches or currant canes and thoroughly drench the plant with any of the winter washes recommended for the San José scale, or the young and recently set scales may be killed by spraying with the contact insecticides recommended for summer treatment of the oyster-shell scale (p. 176).

The Greedy Scale

Aspidiotus rapax Comstock

This European scale insect appeared in California many years ago. It is now abundant there and has gradually spread to the southeast as far as Florida. It attacks various orchard trees, more commonly orange, apple and pear, sometimes appearing on the fruits. The mature female scale is very convex and of a drab or yellowish-brown color, with a dark brown exuvial spot often showing at one side of the center. In California, this greedy scale may be found in all stages at almost any time of the year, even hibernating as eggs, adult females or young. The number of generations annually has not been determined. Four birds, the myrtle and Audubon warblers, wren-tit and bush-tit, eat this scale, their stomachs sometimes being filled with it.

The winter washes recommended for the San José scale will usually control this greedy scale.

The Apple Leaf-hopper

Empoasca mali Le Baron

This is considered the worst all-round leaf-hopper pest, as it works in injurious numbers on so many different plants. Swarms of the active little creatures may attack the foliage of apple, currant, gooseberry, raspberry, potato, sugar beets, beans and celery; more than a dozen other plants, including weeds, grasses, grains and shade trees, are also among its food-plants.

Puncturing the tissues with their tiny beaks, these leaf-hoppers suck the juice, giving the leaves a peculiar, mottled, yellowish appearance and finally causing them to curl. Nursery trees, especially apples, are often seriously injured, the insects working mostly on the undersides of the leaves. The adult insects are about $\frac{1}{8}$ of an inch long and of a pale yellowish-green color with 6 or 8 distinguishing white spots on the front margin of the pronotum (Fig. 178). When disturbed the pale green young or nymphs run in all directions, but the adults can jump quickly

Fig. 178. — The apple leaf-hopper, adult (× 11).

and fly away. The apple leaf-hopper hibernates in both the egg and the adult stages. The winter eggs are deposited in the bark of the smaller branches just below the epidermis, two-year-old wood being most often selected. The position of the

Fig. 179. — Fifth stage nymph of the apple leaf-hopper. Enlarged.

egg is indicated by a low blister-like elevation of the bark about $\frac{1}{30}$ of an inch in length and half as wide. The egg itself is white, elongate, slightly curved and is about $\frac{1}{40}$ of an inch in length. The winter eggs hatch soon after the leaf buds burst in the spring and the hibernating adults appear on the trees about the same time. The young hoppers pass through five nymphal stages and acquire wings at the fifth molt (Fig. 179), about a month after hatching. The summer eggs are not inserted in the bark, but in the petiole and larger veins of the leaves. There are four generations annually in the latitude of Iowa. The first generation works on the lower leaves of nursery trees, doing little real injury, but the later broods, feeding on the tender terminal leaves of the growing shoots in the latter part of June, July and August, seriously retard the growth and thus cause the production of stunted, undersized trees. These hoppers are also an important factor in the dissemination of fire blight among nursery trees.

Remedial treatment.

These leaf-hoppers can be most effectively combated during their nymphal stages, for the adults are so well protected by their wings that applications strong enough to kill them usually injure the foliage. A 10 per cent kerosene emulsion or a solution of whale-oil or any good soap (1 pound in 6 or 8 gallons of water) will kill all the nymphs that are thoroughly hit. Efficient work can also be done with " Black Leaf 40 " tobacco extract, one pint to 100 gallons of water. In the case of nursery stock it is practically impossible to hit enough of the young hoppers in the curled leaves to pay for the labor and expense involved. Dipping the infested tips into pails or dippers containing a soap solution, one pound in 8 gallons of water, kills practically all the young hoppers and is an entirely satisfactory method of controlling the insect on apple nursery stock. The dipping should be done in the latter part of June and again about a month later. At these times the maximum number of nymphs will be found on the trees.

Many of the adults can be captured as they jump and fly away from the disturbed plants by holding near by a shield covered with some sticky tanglefoot-like substance, 1 pound of melted resin in 1 pint of castor oil or " castorine." In Missouri nurseries the hoppers are sometimes captured on sticky shields mounted on a two-wheeled cart drawn by a horse.

REFERENCES

Minn. Agr. Exp. Sta. Bull. 112, pp. 145–164.   1908.
Iowa Agr. Exp. Sta. Bull. 111.   1910.

Bird's Apple Leaf-hopper

Empoasca flavescens variety *birdii* Goding

This leaf-hopper appeared in injurious numbers on apple in Illinois about 1889, and also attacked hops, beans, weeds and walnut trees. Adults were found from May until after heavy frosts. They differ from *Empoasca mali* in often being brighter yellow in color and in having not more than 3 white spots on the pronotum. *Birdii* is considered a color variety of Fabricius' *flavescens* and is distinguished by a smoky band which crosses the middle of the elytra.

CHAPTER V

APPLE INSECTS — BORERS AND MISCELLANEOUS

THE APPLE BUD-BORER

Epinotia pyricolana Murtfeldt

THE cream or pinkish colored caterpillars, about $\frac{1}{4}$ of an inch long, of this Tortricid moth attack young apple trees both in orchards and nurseries, and often the water-sprouts on old apple trees. They mine through the opening terminal buds and continue boring down the twig for an inch or two. Later side shoots are also attacked, and it is often necessary to re-bud trees being top-worked by budding. The insect has been quite destructive in Missouri, Delaware, Maryland and Virginia, and as it occurs in nurseries may spread to other states.

As its whole life-cycle occupies only about six weeks, there are probably four broods annually in its southern range, the second and third broods doing the most damage and its work being the most noticeable in August. The ends of the twigs are killed and a leaf petiole often remains attached to the tips of infested twigs through the winter, thus indicating the presence of the insect, which usually hibernates as a full-grown caterpillar in its burrow, but occasionally in a silken hibernaculum or case covered with bits of bark and dirt on the trunk or branches of the tree. The caterpillars transform in May, and the little, bluish-gray moths prettily marked with brown bands and white dashes and having a wing-expanse of about $\frac{1}{2}$ an inch, emerge and lay eggs from which come a brood of the bud-borers in June. In July and September the second and third broods of the borers get in their destructive work, and probably a fourth brood appears in October.

Observations in Arkansas indicate that in the Ozark region part of the first brood larvæ attack the fruit, that the remainder of the first and nearly all the second brood attack the twigs and water-sprouts and that the larger part of the later broods go back to the fruit. Its work on the fruit is similar to that of the codlin-moth. Many of the caterpillars of the later broods are killed by parasites, and sometimes 50 per cent of those in hibernation are parasitized by *Bracon mellitor*.

A good preventive measure in young orchards or nurseries is to remove the water sprouts from old apple trees near by, as the insect often breeds in these sprouts. As described above, the clinging leaf petioles will reveal many of the infested twigs in winter, and by a judicious pruning and burning of these and other terminals, most of the hibernating caterpillars can be destroyed.

REFERENCES

Del. Agr. Exp. Sta. 12th Rept. for 1900, pp. 194–199.   1901.

U. S. Bur. Ent. Bull. 80, Pt. III, p. 46.   1909.

THE ROUND-HEADED APPLE-TREE BORER

Saperda candida Fabricius

This native American borer ranks among the most destructive enemies to apple and quince culture throughout the northern and eastern portions of the United States and also in Canada. It also occurs in restricted localities in the Southern states. Pear trees are sometimes attacked and its wild food plants include crab apples and thorns of different species, mountain ash, choke cherry, Juneberry and most of the kinds of trees and shrubs in the family Pomaceæ. Although present in most apple and quince orchards throughout the infested territory, it often occurs in injurious numbers in quite restricted localities or spots. We have seen one quince orchard ruined by this borer, while another orchard about two miles away owned and cared for by the same fruit-grower was never seriously infested.

FIG. 180. — Round-headed apple-tree borer, full-grown larva (× 2).

The presence of this borer is usually easily detected at the base of the tree in the spring by the little piles of saw-dust-like castings thrown out from an opening through the bark into its burrow. Often several of the borers work in a tree and the whole tree has a weak and sickly appearance with leaves small and yellowish. Trees of all ages, from nursery stock to large orchard trees, are attacked and often killed. Rank vegetable growths of weeds, grass and water-sprouts around the trunks of trees often seem to afford more favorable conditions for this pest. The borers work mostly in the base of the trunk often below the surface of the ground and in the large roots.

Sometimes they infest the upper portions of the trunk and rarely the larger limbs. The burrows or tunnels begin in the bark and sapwood, but soon extend for several inches up and down in the solid wood, often reaching the heart of small trees.

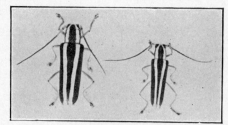

FIG. 181. — Round-headed apple-tree borer, female and male beetles.

The borer when full grown is a light yellow, legless, fleshy grub about an inch in length with a dark brown head and blackish mandibles (Fig. 180). The first thoracic segment is broader than the rest of the body and bears a large patch of many small brownish tubercles on the dorsum and a smaller patch on the ventral side. The constrictions between the segments of the body are deep and the elevated dorsal and ventral portions of the first seven abdominal segments are roughened. The adult insect or parent of this borer is a handsome beetle measuring about an inch in length, the male beetle being considerably slenderer and shorter than the female (Fig. 181). The whole insect, appendages included, is clothed in a velvet-like coating

FIG. 182. — Round-headed apple-tree borer in its burrow in a small apple tree.

of fine, smoothly-laid hairs, giving it a very neat appearance. The long antennæ nearly as long as the body, and the legs are gray. The head and ventral portion of the body are a beautiful silvery white, and from the white face of the beetle two broad, white stripes extend horizontally backward over the head, across the thorax, and along each wing-cover to the tip. The general color of the wing-covers and dorsal portion of thorax is light brown, and the blackish eyes are very conspicuous on the white head.

The round holes (Fig. 184) nearly as large as a lead pencil in the base of the trunk of trees infested by this round-headed borer are the exit holes of the beetles which have developed from the grubs or borers. The beetles emerge mostly at night and remain hidden and inactive during the day. Even in northern localities some of them emerge in April, many of them in May and June, and there are records of their emergence in different localities during the next two or three months. Probably most of the eggs are laid in June, but oviposition may continue to September even in the same locality. The smooth-shelled, pale, rust-brown egg measures an eighth of an inch in length, by one third as wide and is slightly compressed. The eggs are laid in the bark, usually near the ground. The female beetle first makes an incision or slit in the bark, probably with her sharp, horny jaws, but one observer says it is made with the ovipositor. The egg is deposited in the incision, sometimes at the bottom next to the wood, but generally in an opening made in one side of the cut halfway through the bark, nearly a quarter of an inch from the cut. It is then covered with a gummy fluid that sometimes fills the slit and hides the egg, but some observers report that the eggs are easily found. From eggs laid June 15, larvæ hatched in about three weeks in New Hampshire.

The young larvæ soon tunnel through the bark (Fig. 182) to the sapwood in which they work for a year or more, often extending their shallow flat burrows downward below the surface of the ground and remaining dormant in winter. The borers begin work early in the spring, often in March or April, and during the second year of their growth they extend their burrows farther into the solid wood, sometimes going through and girdling young trees. The tunnels often extend upward and downward at various angles in the tree for several inches. Most of the sawdust-like excrement of the grub is packed in its burrow, but some of it is pushed out through small holes eaten through the bark and it often accumulates in little piles at the base of the tree. It is generally believed that it requires three years for this apple-borer to complete its life-cycle. During the third summer the grub sinks its tunnel deeper into the wood, enlarges it and finally extends it outward to the bark. Behind itself the grub packs the tunnel full of sawdust and coarse, woody fibers, and the outward end near the bark is also similarly

Fig. 183. — Round-headed apple-tree borer pupa in its burrow.

plugged. On smooth-barked trees a small, slightly sunken area of dead bark often marks the end of this tunnel, and thus enables one to locate the borer from the outside before it transforms and emerges. A small chamber an inch or more in length is left near the end of the tunnel where the grub hibernates practically secure from the entrance of enemies in either direction. Doubtless some of the grubs do not make this final preparation for their further development into the adult insect until the third spring. Early the third spring, often in May, the grub or borer sheds its skin in this specially prepared chamber and appears in the pupal stage, a delicate, yellowish-white object, somewhat resembling the adult insect (Fig. 183). This

pupa gradually grows darker in color and in about three weeks transforms to the beautiful adult or beetle which soon emerges from a round hole (Fig. 184) about the size of a lead pencil cut through the bark with its strong, sharp mandibles.

As it spends most of the three years of its life inside the tree, this apple-borer has few natural enemies. Woodpeckers get some of the grubs, and at least one

FIG. 184. — Exit hole of the round-headed apple-tree borer.

hymenopterous parasite, *Cenocœlius populator*, sometimes helps to reduce their numbers.

Preventive and remedial measures.

Borers are among the most difficult to control of all insects attacking fruit-trees. No thoroughly satisfactory method of preventing the ravages of this round-headed apple-borer has yet been found. The owner of the quince orchard mentioned in the first paragraph of the discussion of this borer thoroughly tested the commonly recommended preventive washes and shields besides constantly practicing the "digging-out method." Yet so severe was the infestation that all measures failed and the orchard was ruined, the owner giving up the fight and destroying the trees. The statement often made that paper protectors not only prevent the borers from getting into the tree, but also the beetles from emerging, was refuted in this orchard, where the beetles sometimes emerged safely through collars of brick mortar more than a quarter of an inch thick.

In spite of the failure to control the pest in this unusually severely infested orchard, the experiment did demonstrate that certain preventive or deterrent measures were of value. Kerosene emulsion sprayed upon the trunks of the trees several times during the summer seemed to be quite an effective deterrent. One of the best preventives tried was tarred paper closely wrapped around the tree from the roots to a foot or more above ground and well tied on, especially at the top. Alkaline washes of any kind of soap made into a thick paint with a solution of caustic potash or washing soda and about 1 pint of crude carbolic acid added to every 10 gallons of the wash have often been used as deterrents with good results. Two or three applications should be made from early May to July, thoroughly covering the trunk from the branches down to the roots. Instead of tarred paper, old newspapers or wire mosquito netting may be used as mechanical protectors. Let the netting loosely encircle the tree except at the top so the beetles cannot reach the bark of the tree through the meshes. These open wire protectors will last two or more seasons, while the close paper ones should be removed in the fall. Some orchardists make a mound of earth from five to twelve inches high around the base of the trees in early spring, thus forcing the beetles to lay their eggs higher up on the trunk where the grubs can be more easily found and dug out. The mounds should be removed in early autumn. A combination of this mounding system or a wire or paper protector with the applications of a deterrent wash above the protector if thoroughly done should effectually prevent many of the borers from getting into the trees.

Clean culture is one of the best preventive measures. Do

not allow rank growths of weeds, water-sprouts or other vegetation to accumulate about the base of the tree.

The surest and best remedy for this pest after it gets into the tree is the old and much practiced digging or cutting out method. Experienced orchardists are often able to readily locate the borer by the oozing of sap or by particles of sawdust coming from minute holes in the bark leading into the burrows. It is then often an easy matter to reach the borer with a sharp knife or chisel, or a wire may be pushed into the tunnel until the grub is impaled. Great care must be practiced or more injury may result from the knife than from the borer. The trees should be thoroughly examined in early spring, not later than May, for the borers are usually more easily located then by their sawdust-like castings, and it is also important to destroy those that are then transforming to the beetles. Continue to use the knife during the summer whenever a borer can be located, and go over the trees thoroughly in early autumn to get the younger borers working just beneath the bark. Where valuable trees have been riddled or girdled by the borers, or in trying to dig them out, the tree may be saved by putting in several bridge grafts at the base, as is often done when trees are girdled by mice or rabbits. Wounds made in removing borers should be coated with gas tar to exclude moisture and prevent the development of fungous diseases.

Other kinds of borers in shade or park trees have been successfully treated by injecting a little carbon bisulfide into the small hole from which the sawdust-like excrement is being pushed out, and the hole quickly plugged with putty or grafting wax. The deadly fumes of this very volatile liquid penetrate the burrow and finally kill the borers. Orchardists should give this sensible method a thorough trial. A similar treatment has been successfully used by some fruit-growers. Wherever the sawdust is seen coming through the bark, kerosene is freely applied which is absorbed by the castings and carried by capillary attraction through the burrow, finally coming in contact with and killing the grub. The small amount of kerosene necessary to accomplish this is said not to injure the tree. Do not waste time and materials in trying to reach the borer through the large round holes, as large as a lead pencil, for these are the exit or emerging holes of the adult insect or beetle. The grub or borer has finished its nefarious work and transformed into the handsome beetle, which made the hole, and flew away to seek its mate and provide for more destructive work by their progeny.

REFERENCE

U. S. Bur. Ent. Circ. 32 (third revise). 1907.

THE SPOTTED APPLE-TREE BORER

Saperda cretata Newman

This Cerambycid beetle is very similar to its near relative, the round-headed apple-borer, both in appearance and habits. The beetle of the spotted borer is about the same size and form, but differs in being of a darker brown color with its legs, antennæ, head and the middle portion of the ventral surface of the same brown color. Two broad, silvery white stripes extend along the sides of the thorax and abdomen and there are two similar narrow stripes on the dorsum of the thorax. The continuous white stripes on the wing-covers of the beetle of the round-headed borer are replaced by two large white spots on the

FIG. 185. — The spotted apple-tree borer ($\times 2\frac{1}{2}$).

o

wing-covers of this borer (Fig. 185). Although this insect is widely distributed throughout practically the same territory as its near relative, it has been recorded as injurious only in Iowa and Michigan. Besides injuring apple and wild crab trees, it also attacks Juneberry and thorn. The beetles are said to lay their eggs in the bark in pairs, half an inch or more apart. The grubs of each pair upon hatching then work in opposite directions around the trunk or branch, at first just beneath the bark, and afterwards entering the hard wood.

The remedial measures suggested for the round-headed borer, *Saperda candida*, will also apply to this spotted borer, except that the latter often works in the larger branches as well as the trunk, thus rendering it necessary to extend the protective or preventive treatments to the branches.

REFERENCE

U. S. Bur. Ent. Circ. 32 (third revise). 1907.

The larva of another long-horned beetle (*Leptostylus aculiferus* Say) sometimes burrows under the bark of diseased apple trees. The beetle is about $\frac{1}{3}$ inch in length, and brownish-gray in color. The wing-covers are ornamented with numerous small thorn-like points and are crossed, behind the middle, with a V-shaped band, margined with black. The adults may be found from August to September.

THE FLAT-HEADED APPLE-TREE BORER

Chrysobothris femorata Fabricius

This Buprestid beetle is widely distributed throughout the United States and southern Canada. It is often found basking in the sunshine on fallen trees and the warm, sunny sides of the trunks of many kinds of trees, but the shy creature is not easily captured, as it runs rapidly or flies readily when approached. The favorite natural host-plants of the insect are doubtless oak trees, but it attacks a great variety of wild and cultivated plants. Among orchard fruits it is often injurious to apple, quince, pear, peach, plum and apricot trees and currant bushes. Pecan trees are attacked by it in Alabama, and its food-plants among shade and forest trees now include oak, mountain ash, maple, box-elder, hickory, chestnut, sycamore, horse-chestnut, linden and willow.

The common name of this insect, the *flat-headed* borer, was suggested by the peculiar flat-headed appearance of the larva or grub to distinguish it from the *round-headed* borer, *Saperda candida*, often working on the same trees. While the flat-headed borer is more common, it is usually a less dangerous pest than the latter species, because it rarely attacks thrifty, healthy trees. It seems to prefer the warmer southern sides of young trees that are suffering from lack of cultivation, uncongenial soil, climatic or other disease-producing conditions.

FIG. 186. — Adult of the flat-headed apple-tree borer ($\times 3\frac{1}{2}$).

This destructive short-horned metallic beetle is of a flattish oblong form (Fig. 186) and about half an inch long, varying considerably in size. The antennæ are short, the eyes large, and the front legs are armed with a conspicuous tooth. The upper surface of the body is of a dark, coppery-brown color and fresh specimens are often coated in spots with a powdery gray substance that easily rubs off. Each wing-cover has three slightly raised lines, the outer two of which are interrupted by two impressed, irregular, transverse spots of a brassy green color, dividing each wing-cover into three nearly equal portions. The under surface of the body and the legs are of a bright metallic greenish-blue, shining like burnished copper when the insect is flying. The males are smaller than the females and have shining green heads. These sun-loving beetles begin to appear early in May and continue through July even into September. The females deposit their yellow, irregularly ribbed eggs, about $\frac{1}{50}$ of an inch in length, in cracks or under bark scales, usually several eggs in a place, either on the trunk or branches, and almost always on the warmer southern side of the tree, or on recently felled logs in sunny locations. The larva or grub which hatches from these eggs soon eats its

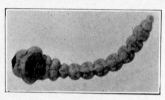

way through the bark and excavates a broad, flat, irregular channel, often extending into the sapwood just under the bark. Part of the channel is packed with sawdust-like castings of the grubs.

FIG. 187. — Flat-headed apple-tree borer ($\times 1\frac{3}{4}$).

A single borer may often girdle the trunk and kill small trees. The full-grown, light-yellow, legless grub is about an inch long with its second thoracic segment much broadened and flattened (Fig. 187). On the upper and lower surfaces of this segment are large, roughened spots with two smooth diverging linear depressions extending through the dorsal spot and one medially through the ventral spot. The grub habitually rests in a curved position and gets its full growth during a single summer. It finally extends its burrow outward nearly through the bark, then digs a little deeper into the solid wood, forming a chamber in which it transforms through the tender white pupal stage (Fig. 188) to the adult insect. In the north these pupal chambers are made in the spring, the winter months being passed as grubs, but farther south pupation may occur in November, the pupæ hibernating. The pupal period lasts about three weeks in the spring, and the beetles cut their way out through an elliptical hole. Thus this flat-headed borer has a much shorter life-cycle than the round-headed species, from which it also differs much in all its stages and in its habits.

Although snugly hidden in its burrow beneath the bark, this flat-headed borer does not escape from natural enemies. Woodpeckers and ants devour many of the grubs and pupæ and the following parasites prey upon it: the Braconids, *Bracon charus* and *pectinator, Spathius pallidus;* the Ichneumonids, *Labena apicalis* and *grallator;* and one or more species of Chalcids.

Fig. 188. — Pupa of the flat-headed apple-tree borer.

Remedial suggestions.

As healthy, well-established trees are rarely attacked by this borer, its ravages can be largely prevented by keeping young trees in a thrifty, growing condition. Many of the suggestions for controlling the round-headed borer are also applicable to this species. The deterrent soap and emulsion washes and protective coverings have been found effective, but must be applied farther up the trunk and on to the larger branches. Two or more applications of the washes should be made, beginning in May and continuing into July. As the location of the larval channels are often revealed by slight discolorations or diseased conditions of the bark or by sawdust-like castings thrown out of the burrows, the culprit can often be easily reached with a sharp tool. Several of the grubs often infest a young tree, completely undermining the bark over considerable areas; we once found on a small apricot tree three pupæ in an area not larger than a penny. In cases of such severe infestation the trees should be removed and burned in autumn or before May.

A closely related Buprestid beetle, *Chrysobothris mali* Horn, which has been given the common name of "California big-headed borer" has been recorded as injurious to apple trees in Arizona. If its life habits are similar to those of *Chrysobothris femorata* just discussed, the same remedial measures should prove effective in controlling it.

REFERENCE

U. S. Bur. Ent. Circ. 32 (third revise). 1907.

THE APPLE WOOD-STAINER

Pterocyclon (Monarthrum) mali Fitch

Half a century ago this minute, reddish-brown Scolytid beetle (Fig. 189), only about $\frac{1}{10}$ of an inch in length, was

FIG. 189. — The apple wood-stainer beetle (× 14).

reported from Massachusetts as riddling the trunks of apple trees with their burrows. The insect belongs to the interesting group of beetles known as Ambrosia beetles, which propagate a mold-like fungus in their burrows that stains the walls black, and is eaten by the beetles and fed by them to their offspring or grubs. The parent beetle bores through the bark into the wood for about $\frac{1}{4}$ of an inch, then excavates a main transverse tunnel or gallery in the solid wood from the sides of which short galleries made and occupied by grubs extend at right angles upward or downward. This wood-stainer breeds only in diseased or dying, felled or girdled trees, and sawed mahogany lumber. Its food-plants include nearly twenty different forest or timber trees, as well as the apple, orange and morello cherry among fruit-trees. It has also been called the Lesser Cask-beetle because of its fondness for boring into wine, beer and vinegar casks, often doing much damage in this way. As there are no recent records of its injuring orchard trees, it needs no further discussion as a fruit pest. The deterrent washes recommended for the larger apple-borers will doubtless prove effective should the insect again appear in injurious numbers in orchards.

The Bronze Apple-tree Weevil

Magdalis œnescens Leconte

In Oregon, Washington, Montana and British Columbia apple trees, particularly young trees, are sometimes attacked and killed by this small snout-beetle or weevil (Fig. 190). Weak or sickly trees are more liable to attack. The insect may work on the trunk, but more often makes its tunnels under the bark of the branches and may continue its work after the trees are dead. The little, plump, legless, white grubs

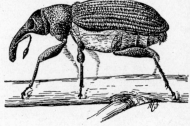

FIG. 190. — The bronze apple-tree weevil (× 10).

about $\frac{1}{6}$ of an inch in length may extend their narrow burrows for an inch or two in various directions under the bark. Each burrow ends in a little cell where the grub transforms, probably in the spring, through the pupal stage to the adult or weevil. There is apparently but one generation of the insect in a year, the weevils emerging through small, round holes in the bark from early April until August, and laying their smooth, shining, yellowish-white eggs in little horizontal holes dug in the bark. The slender, blackish-bronze colored weevils, measuring about $\frac{1}{6}$ of an inch in length, often feed upon the leaves of apple and cherry, sometimes nearly defoliating small trees. The wild food-plant of the insect seems to be a species of thorn.

Two Chalcid parasites destroy large numbers of the grubs.

The repellent soap and lye washes or the kerosene emulsion recommended for the round-headed apple-tree borer have been found effective against this bronze apple weevil if an application is made early in April and repeated late in May. Badly infested trees that are dying should be removed and burned before the weevils begin to emerge in April.

REFERENCE

U. S. Bur. Ent. Bull. 22, pp. 37–44. 1900.

THE TWIG-PRUNER

Elaphidion villosum Fabricius

In early autumn an orchardist's attention may be attracted to twigs or branches from a few inches to two or three feet in length which have fallen to the ground, having been deftly pruned from the trees by this twig-pruner. An examination of the severed end of the branch will reveal a smoothly cut surface, near the center of which is an oval opening plugged with a wad of fine shavings and sawdust. If the branch be split open, it will usually be found to have been tunneled, often from 10 to 15 inches, by a slender, whitish grub nearly $\frac{3}{4}$ of an inch in length, then lying in the tunnel near the severed end, or the insect may have already reached the pupal or even the adult or beetle stage. The grub is quite hairy, and it has three pairs of minute legs. Usually the grubs transform in the spring and the beetles continue to emerge from June till September. This Cerambycid beetle is a slender, reddish-brown insect rather sparsely covered with yellowish, somewhat mottled pubescence. It varies considerably in shape and size, measuring from $\frac{1}{2}$ to $\frac{3}{4}$ of an inch in length with the antennæ of the female a little shorter and those of the male longer than the body. The proximal joints of the antennæ are armed with small spines, and each wing-cover terminates in two small spines.

The eggs are inserted in the smaller twigs of living trees in summer. The young grub hatching therefrom feeds for a time on the softer tissues under the bark, packing its burrow with its sawdust-like castings and gradually boring towards the base of the twig. As the borer grows it often consumes the larger portion of the wood and ejects some of its castings through holes made in the bark. Later it bores along the center, making a more or less oval chamber. Early in the fall it eats away nearly all the wood, plugs the end of its burrow and waits for the wind to break off the nearly severed branch. The purpose of the grub in this pruning operation is not definitely known. Early writers attributed to the grubs an unwarranted degree of intelligence, almost reasoning powers, by which they were considered infallible, knowing just how far to cut and being able to vary the operation to meet the circumstances in each particular case. The most plausible explanation yet suggested is that the grub nearly severs the branches and plugs the free end of the burrow to provide for the emergence of the weak-jawed beetles, which could never eat their way out through the solid wood. Under natural conditions in the field the life-cycle of this twig-pruner is doubtless completed in a year, but when infested twigs are gathered and kept in dry situations, it may require three or four years to breed the beetles.

Oak and maple are favorite food-plants of this insect and the ground is often strewn with severed branches under these trees. It attacks many other forest or shade trees and shrubs, and the following fruits are also pruned by it, sometimes quite severely : apple, pear, quince, peach, plum, grape and orange.

Several birds, woodpeckers, blue jays and chickadees destroy many of the grubs or pupæ in the fallen twigs. A parasite, *Bracon eurygaster*, has been reared from infested twigs.

The collection and burning of the fallen branches in autumn or early spring will effectually control the pest.

REFERENCE

U. S. Bur. Ent. Circ. 130. 1910.

The Twig-girdler

Oncideres cingulata Say

Twigs and branches less than half an inch in diameter on many kinds of forest and shade trees and on several of the orchard fruits are often neatly girdled by a handsome, robust, ash-sprinkled, reddish-brown beetle a little more than half an inch in length with antennæ longer than its body and a broad, ashy-colored belt around the middle of the wing-covers and across the thorax; closer inspection also reveals numerous

Fig. 191. — The twig-girdler.

light brown spots on the wing-covers. The beetles appear during July and August and the girdling is done by the females standing on the twig head downwards and cutting the girdle section by section about an eighth of an inch wide and extending to the heartwood, so that the branch is easily broken off by high winds (Fig. 191). During the girdling process, which often occupies half a day, the female stops several times to move outward on the twig and tuck an egg underneath the bark at the base of a side shoot or an aborted bud. The girdled twigs are soon broken off and fall to the ground, where most of the eggs hatch by autumn. In the spring the grubs bore into the solid wood and often make a channel 2 inches long and disposing of nearly all the woody portion of the twig, but always leaving the bark intact. The white, legless grubs about $\frac{3}{4}$ of an inch long have a row of short, parallel, chitinous ridges near the front margin of the head, a portion of the dorsum of the thorax is striated and elevated, and there are double, transverse rows of minute, brownish, toothlike projections both on the dorsal and ventral portions of the third to the tenth segments. Early in July these grubs block up the ends of their channels and all holes or cracks in the bark with small shavings and at one end transform through the pupal stage to the adult insect in about two weeks.

This twig-girdler is common throughout the eastern and southern portions of the United States, and while working mostly on elm, hickory and other forest trees, it includes the following fruit-trees among its food-plants : apple, pear, quince, peach, plum, cherry and persimmon. The beetles are the depredators, their progeny doing no harm by working in the dead twigs on the ground. Sometimes the ground under shade trees is strewn with the girdled twigs, but fruit-trees are rarely seriously injured.

The remedy is simple. Gather and burn the fallen twigs in autumn, winter or early spring, thus destroying the eggs and young grubs in their hibernating quarters.

REFERENCES

Kan. Agr. Exp. Sta. Bull. 77, pp. 56–62. 1898.
Okl. Agr. Exp. Sta. Bull. 91. 1911.

FLEA-BEETLES

At least half a dozen different kinds of these small, active beetles, with enlarged hind legs that enable them to jump like fleas, sometimes attack and seriously injure young apple trees, especially newly budded or grafted nursery stock and seedlings. The beetles swarm from near-by weeds upon the trees in May and June and eat small holes through the leaves, often riddling the foliage and causing the death of young grafts or seedlings. Most of the species attacking apples are common and widely distributed over the United States and Canada.

The apple flea-beetles (*Haltica foliacea* Leconte; and *Haltica punctipennis* Leconte)

These two bright, shining green flea-beetles are about $\frac{1}{6}$ of an inch in length and distinguished from each other with difficulty. The first species lives mostly in the southwest, from Texas to Arizona, while *punctipennis* usually has more distinctly punctuated wing-covers and is common from Missouri westward to California. Most of the injuries to apple trees, although recorded as the work of *foliacea*, are now considered to have been done by *punctipennis*. Injury to young apple trees by these two species has been reported mostly from Kansas, Colorado and New Mexico. They are the only species of these flea-beetles attacking apple which are known to breed upon the apple, their blackish grubs sometimes skeletonizing the leaves in July. There is apparently but one generation annually, the beetles hibernating.

The pale-striped flea-beetle (*Systena tœniata* Say)

This flea-beetle, about $\frac{1}{8}$ of an inch long, varies in color from almost black to a pale brownish, and it has a whitish longitudinal stripe extending down the middle of each wing-cover. Its grubs are root-feeders on weeds and corn, but the beetles are very general feeders. The terminal buds, and later the leaves, on apple and pear grafts have been destroyed by the beetles in Arkansas, and 20,000 seedling apples were stripped in seven days and killed in New York in June, 1900.

The smartweed flea-beetle (*Systena hudsonias* Förster)

This totally black flea-beetle, about $\frac{1}{5}$ of an inch long, generally feeds on smartweed, dock and other weeds, and little is known of its life history. In June and July, 1897, it swarmed on to apple and pear grafts and 2-year-old trees in a New York nursery, and fed voraciously on the upper and lower surfaces of the leaves, riddling them and killing many of the grafts.

The willow flea-beetle (*Crepidodera helxines* Linnæus)

This tiny, European flea-beetle, only $\frac{1}{10}$ of an inch in length, is smooth, and varies from a brownish bronze to a metallic blue or green color. It is often found on willows eating small, round holes in the leaves, and the beetles hibernate, appearing in May and June. They once riddled the foliage of young apple trees in an Illinois nursery in June. The cucumber flea-beetle (*Epitrix cucumeris* Harris) was also detected at the same work in Illinois early in May. It is about $\frac{1}{3}$ smaller than the willow species, of a shining black color, somewhat hairy, and is a common pest in gardens on potatoes, cucumbers and other vines. Its grubs sometimes cause "pimply" potatoes.

Another Chrysomelid beetle (*Syneta albida* Leconte), although not of the group known as Flea-beetles, attacked apple, cherry and peach trees in a similar manner in Oregon in 1892. The slender, yellowish-brown (the females are yellowish-white) beetles, about $\frac{1}{4}$ of an inch in length, riddled the foliage and blossoms in April, sometimes cutting half through the stems of the young fruits. A sudden jar of the infested trees caused many of them to drop, so they could be readily collected on sheets or curculio-catchers, or a thorough application of a poison spray would kill them.

The red-legged flea-beetle (*Crepidodera rufipes* Linnæus)

This little flea-beetle, about $\frac{1}{10}$ of an inch long, with shiny, blue wing-covers and reddish-brown head, antennæ, thorax and legs, is common in Europe and widely distributed in the United States, where its favorite food seems to be black locust foliage. Young apple, pear, peach and plum orchards set near or on locust clearings have been seriously injured in early spring by the beetles swarming on to the trees and eating out the buds, sometimes denuding and killing the trees. It is suspected that the insect breeds on locust roots, but its life history is unknown, except that the beetles hibernate in rubbish on the ground. Thorough and frequent spraying of the opening buds with arsenate of lead (6 or 8 pounds in 100 gallons) will check the ravages of the beetles, which may also be jarred from the trees on to sheets or plum curculio-catchers. Kerosene emulsion diluted with 6 or 8 parts water is said to kill all that are thoroughly hit.

Remedies for flea-beetles.

Experiments have shown that these flea-beetles can be effectually controlled and many of them killed by one or two thorough applications of a strong poison spray, as Paris green, 1 pound in 100 gallons of water, or arsenate of lead, 4 or 5 pounds in 100 gallons of water. Bordeaux mixture alone often protects the foliage from their attacks, and, combined with a poison, will make a very effective spray. Apply as soon as the beetles appear and make a second application a few days later, if necessary.

THE CLOVER-MITE

Bryobia pratensis Garman

This is a minute, spider-like, oval-shaped, reddish-brown mite about $\frac{3}{100}$ of an inch in length and with remarkably long front legs (Fig. 192). Although it is closely related to the common, two-spotted mite or so-called "red-spider" of greenhouses, it lives mostly out-of-doors on trees, clover and grasses. Throughout Canada and the northern half of the United States, especially during dry seasons, this clover-mite often swarms over the foliage of peach, prune, plum, apple, pear, cherry, almond, raspberry and other fruits, and many forest trees also. It is one of the principal enemies of fruit-trees in the far western orchard sections. These mites have mouth parts fitted for piercing the tissues, and badly infested foliage assumes a blanched, yellowish, sickly appearance.

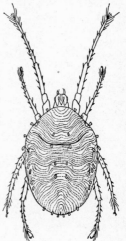

FIG. 192. — The clover-mite. Redrawn after M. A. Palmer (× 37).

In warmer localities many of the mites hibernate on the bark or in other sheltered locations, but in northern regions the winter is passed in the egg stage. Oftentimes the bark of fruit-trees, especially in the crotches, is covered in winter by thousands of the tiny, round, reddish eggs (Fig. 193). The conspicuous, rusty, reddish appearance given to the bark by the masses of mite eggs often first leads to the discovery of the pest. The eggs hatch early in the spring and breeding continues through several generations during the growing season. In autumn housewives are often much annoyed by swarms of the mites coming from near-by clover fields and seeking hibernation quarters in the house.

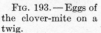

FIG. 193. — Eggs of the clover-mite on a twig.

REFERENCE

U. S. Bur. Ent. Circ. 158. 1912.

The two-spotted mite (*Tetranychus bimaculatus* Harvey)

This is the common "red-spider" of greenhouses and the mite often discussed as *Tetranychus telarius*. Under this latter name there are many reports of injury to fruit-trees by "red-spiders," but in most cases it is evident that the real culprit was the clover-mite. The two-spotted mite is smaller, about $\frac{1}{50}$ of an inch long, and the legs are more nearly of equal length. The body and legs bear many stout hairs; it varies in color from yellow through orange to brown and dark green, often with a darker spot on each side of the body. It spins a very delicate silken web-like nest over its breeding grounds. Its eggs are elongate and white, and it is said to hibernate among fallen leaves and not as eggs on the bark. Thus while these two mites work in a similar manner on the foliage, they differ considerably in appearance and life-habits. (See also p. 315.)

Remedial treatment for these mites.

The clover-mite can be most effectively and easily controlled by thoroughly spraying the hibernating eggs in fall, winter or early spring with lime-sulfur at the strength usually used against scale, or a 10 per cent kerosene emulsion. The mites can be killed on the foliage with whale-oil soap solution (1 pound in 10 gallons water) or by dusting with sulfur and hydrated lime.

REFERENCES

Vt. Agr. Exp. Sta. 10th Ann. Rept. pp. 75–86. 1897.
Col. Agr. Exp. Sta. Bull. 152. 1909.

THE RING-LEGGED TREE-BUG

Brochymena annulata Fabricius

This large, widely distributed, dull grayish-brown Pentatomid bug about $\frac{5}{8}$ of an inch long often occurs in young apple orchards and has been accused of killing the tender, young shoots by sucking out the sap in May. It is also reported as common on plum, cherry and larch trees and grape vines. The eggs and nymphs have been found on pea vines and willow trees and the old bugs in hibernation under bark. Another species, *Brochymena 4-pustulata*, in both nymphal and adult stages, is recorded as sucking the juices from the pupæ of the white-marked tussock-moth in Delaware, so that this ring-legged tree-bug may also include injurious insects as a portion of its diet, thus offsetting some of the injury it may do to plants.

The adult bugs can be jarred or hand-picked from the trees on to sheets, or the nymphs killed by spraying with kerosene emulsion, whenever the insects become injurious in orchards or vineyards.

The Eye-spotted Apple-twig Borer

Oberea ocellata Haldeman

The new growth or twigs of young apple trees are sometimes infested with a bright-yellow, deeply incised, legless grub about half an inch long when mature and bearing a characteristic shield-shaped, horny, roughened brownish plate arising obliquely from the head and covering the dorsum of the first thoracic segment. These grubs devour the pith and also the woody fiber of the twigs, ejecting their castings through pin-like holes cut in the bark at irregular intervals. There is apparently one generation of the insect in a year, the grubs maturing in the fall and hibernating in their burrows, where they quickly transform through the pupa stage to the adult insect in April and May. The slender Cerambycid beetles, measuring about an eighth of an inch in length, are of a pale, reddish-brown color with their long antennæ, wing-covers and feet very dark brown or blackish. Two and sometimes four conspicuous black spots occur on the dorsum of the thorax. Besides apple twigs, the insect also breeds in the twigs of peach, plum, pear and poison

P

sumac. Although quite widely distributed in the United States and occurring also in Canada, this twig-borer seems to have been injurious only in Missouri and Texas. The only practicable remedy thus far suggested is to prune off and burn all infested twigs at any time before April.

The New York Weevil

Ithycerus noveboracencis Förster

This large, ash-gray, black-spotted beetle (Fig. 194), one of the largest of the weevils, measuring about $\frac{5}{8}$ of an inch in

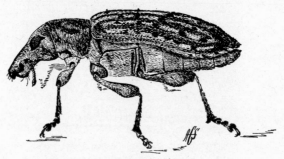

Fig. 194. — The New York weevil ($\times 3\frac{2}{3}$).

length, is widely distributed in the United States and Canada, but has been injurious principally in the Mississippi Valley and the Southern states. It breeds in the twigs of oak, hickory and possibly other forest trees. The weevils appear in early spring and eat into the buds, gnaw the tender bark on new growth, and often cut off the leaf-stalks and new shoots, working principally at night. Young apple, peach, plum, pear and cherry orchards and nursery trees may be invaded by the beetles and many trees ruined.

These large weevils can be readily controlled either by handpicking or by jarring them on to sheets or plum curculio-catchers.

The Snowy Tree-cricket

Œcanthus niveus De Geer

Apple and plum trees, especially if growing in orchards overgrown with weeds and other rank vegetation, are liable to be injured to a slight extent by the egg punctures of this pale yellowish-white tree-cricket. This species (*niveus*) was for-

Fig. 195. — A pair of snowy tree-crickets courting, the male with wings expanded.

merly supposed to deposit its eggs in rows of punctures in raspberry canes, but recent work at the Geneva Experiment Station has shown that another species (*nigricornis*) is the real cause of the injury to raspberry and that *niveus* deposits its eggs singly in punctures in the bark of the smaller branches of apple, peach, plum and other trees (see p. 325). The egg-punctures often permit the entrance of fungous spores and bacteria which cause the surrounding bark to become diseased, and produce discolored areas of dead bark known as cankers, or woolly aphids may start a colony at the wound and cause a bad scar on the branch. Figure 196 shows an egg of the snowy tree-cricket in position in a small apple branch. The egg is laid in late

summer or autumn and hatches the next May or June. The adults of this and closely allied species sometimes injure ripe plums, grapes and peaches by eating out holes about the size of the insect's head. The injury is often attributed to bees, but the latter only collect the juice from the wound made by the tree-cricket.

The injury from tree-cricket punctures is rarely found in orchards kept free from rank vegetation. Where cankers have started to form they should be cleaned out with a sharp knife, cutting back to the live bark, and the wound should then be treated with a coat of gas tar to prevent infection.

FIG. 196. — Section of an apple branch showing egg of snowy tree-cricket in position. Greatly enlarged.

REFERENCE

Parrott, Jour. Ec. Ent. II, pp. 124–127. 1909.

OTHER APPLE INSECTS

PEAR-LEAF BLISTER-MITE: *pear*, p. 227.
PEAR BLIGHT BEETLE: *pear*, p. 232.
HOWARD SCALE: *pear*, p. 234.
EUROPEAN PEAR SCALE: *pear*, p. 234.
PLUM CURCULIO: *plum*, p. 243.
AMERICAN PLUM BORER, p. 253.
EUROPEAN FRUIT-TREE SCALE: *plum*, p. 260.
EUROPEAN FRUIT LECANIUM: *plum*, p. 261.
FRUIT-TREE BARK-BEETLE: *peach*, p. 277.

TERRAPIN SCALE: *peach*, p. 293.
CHERRY SCALE: *cherry*, p. 312.
WALNUT SCALE: *currant*, p. 360.
IMBRICATED SNOUT-BEETLE: *strawberry*, p. 371.
ROSE CHAFER: *grape*, p. 397.
COTTONY MAPLE SCALE: *grape*, p. 427.

CHAPTER VI

PEAR AND QUINCE INSECTS

THE most injurious insects attacking the pear are the codlin-moth (p. 10), San José scale (p. 162), the pear psylla and in some regions the pear thrips. While more distinctly an apple tree pest, the codlin-moth annually causes a loss of nearly a million dollars to the pear crop.

Owing to the fact that the calyx lobes of the pear remain open and do not close up as in the apple, spraying for the codlin-moth is less effective on pears than on apple. Insects often cause serious injury to pears indirectly by distributing the spores of the bacterium, causing the disease known as fire-blight. Their legs and mouth parts become smeared with the sticky liquid containing the spores, which are thus introduced into the tissues of the plant by the claws in walking over the tender tips or by the beak when feeding.

THE PEAR SLUG

Eriocampoides limacina Retzius

For nearly two centuries this insect has been recognized as an enemy of the pear and cherry in Europe, while in this country its history goes back to the extensive account published in 1799 by William D. Peck of Massachusetts. It was probably introduced into New England in colonial times and is now generally distributed wherever its food-plants are grown. While in Europe it has been reported as feeding on a large number of

214

FIG. 197. — Egg-blisters of the pear slug.

plants, in this country its injuries are confined almost entirely to the pear, cherry and plum. Eggs are frequently laid in peach leaves on trees adjoining infested pear and cherry orchards, but the larvæ do not seem to thrive on that food plant.

In the North the small, glossy black, four-winged flies about $\frac{1}{5}$ inch in length appear on the leaves about the middle of May. The female is provided with a sharp saw-edged ovipositor by means of which she deposits her eggs under the epidermis of the leaf. The ovipositor is inserted from the under surface of the leaf and then so manipulated as to cut loose a portion of the upper epidermis,

forming a kind of blister in which the oval egg is laid (Figs. 197 and 198). The egg hatches in about two weeks and the whitish young larva escapes on the upper side of the leaf through a semicircular cut in the overlying epidermis. The larvæ soon become covered with a brownish sticky slime, which is retained until they are full grown and gives them the

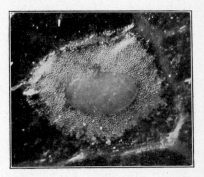

FIG. 198. — Egg-blister of the pear slug, greatly enlarged.

appearance of small snails. The body is swollen in front and tapers behind like a tadpole (Figs. 199 and 200). The larva

Fig. 199. — Pear slugs feeding on a leaf.

passes through five stages; at the fourth molt it loses its slimy covering and in the fifth stage is of a light orange yellow color.

Fig. 200. — Pear slugs feeding, enlarged.

It does not feed in this stage, but crawls or falls to the ground in which, at a depth of two or three inches, it constructs its earthen cocoon. Most of the larvæ transform to pupæ in six to eight days and the adults emerge about ten days later and lay eggs for a second generation, but a few of the first brood larvæ remain unchanged in their cocoons until the following spring.

In New York the eggs for the second brood are laid during late July and early August. Normally there are two broods in the North, but three have been reported in the latitude of Washington, D.C.

The larvæ feed on the upper surface of the leaves eating only the epidermis and leaving the skeleton of veins and the lower epidermis to turn brown and wither up. Badly injured leaves fall and the tree may be entirely defoliated by midsummer; the fruit becomes stunted, fails to mature, and the vitality of the tree is so weakened that fruit buds for the next year's crop are not formed. Sometimes the tree will put out a new crop of leaves, but these in turn may be destroyed by the second brood of slugs. In cases of such severe infestation the orchard has the appearance of having been swept by fire. Fortunately this pest is rarely troublesome for a series of years in the same locality, probably being held in check by a minute egg parasite.

Treatment.

When only a few trees are to be treated the slugs may be destroyed by one or two applications of freshly slaked lime dusted on the leaves. White hellebore, one ounce to three gallons of water, has also given good results. For commercial orchards arsenical sprays are more satisfactory. Arsenate of lead because of its greater safety and effectiveness is preferable to the older arsenicals, Paris green and London purple, and should be used at the rate of 4 pounds to 100 gallons of water. The slugs can also be killed by spraying with "Black Leaf 40" tobacco extract, 1 pint in 100 gallons of water, adding 4 or 5 pounds of soap to make the liquid stick and spread better.

References

Nev. Agr. Exp. Sta. Bull. 10. 1890.
U. S. Bur. Ent. Circ. 26. 1897.
Col. Agr. Exp. Sta. 15th Rept. pp. 11–13. 1903.
Wash. Agr. Exp. Sta. Bull. 65, pp. 12–14. 1904.

THE PEAR PSYLLA

Psylla pyricola Förster

This serious enemy of the pear was introduced into Connecticut from Europe about 1832. It is now generally distributed over the Eastern states and Canada, extending southward to Virginia; it also occurs in California. Locally the abundance of the pest varies greatly from year to year; severe outbreaks lasting over two or three years are usually followed by longer periods of comparative immunity.

Badly infested trees take on a sickly appearance early in the season, the leaves turn brownish or black, dry up and fall in midsummer; the fruit remains small and much of it falls prematurely. Some orchards give the impression of having been swept by fire, and owing to their weakened condition are especially liable to winter killing.

The adults hibernate on the trunks in crevices and under flakes of bark, and when very abundant they may collect under leaves and trash on the ground. The adult psylla has been aptly likened

FIG. 201.— The pear psylla, adult (× 20).

to a diminutive cicada or dog-day harvest-fly (Fig. 201); they are about $\frac{1}{10}$ inch in length, dark reddish-brown in color, with the abdomen banded with black. When at rest the two pairs of large, nearly transparent wings slope roof-like over the sides of the body.

With the first warm days of spring the adults emerge from their winter quarters and egg-laying begins in a few days. The yellowish-orange eggs, about $\frac{1}{80}$ inch in length are deposited in the creases of the bark, in old leaf scars and about the base of the terminal buds. They are elongate pyriform in shape and have a smooth shining surface. A short stalk at the larger end attaches the egg to the bark and a long thread-like process projects from the smaller end. A large proportion of the eggs is laid before the buds open. They hatch in 11 to 30 days, depending on the temperature.

Most of the eggs have hatched by the time the petals fall. The recently hatched nymphs are of a translucent yellow color and are hardly visible to the unaided eye, being scarcely $\frac{1}{80}$ inch in length. They migrate at once to the axils of the leaf petioles and stems of the forming fruit, and when these places become crowded they will scatter out on the underside of the leaves and on the petioles. They are provided with sucking mouth parts and feed exclusively on the sap. The larger part of the sap taken into the body is elaborated into a sweet, sticky substance known as honey-dew, which is voided from the tip of the abdomen and collects as a large glistening drop. The leaves and fruit on infested trees are always more or less covered with this unpleasant secretion and soon acquire a disgusting blackish appearance from a sooty fungus which grows upon it. The presence of the honey-dew is frequently the first indication that the tree is infested.

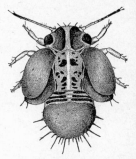

FIG. 202. — The pear psylla, last stage nymph. Enlarged.

The psylla passes through five immature stages and acquires wings at the fifth molt (Fig. 202). About a month is required for the complete life-cycle, and there are at least four broods a season. The females of the later generations do not lay their eggs on the bark, but deposit them along the midrib on the underside of the leaves or place them in the notches at the edge of the leaf. The adults of the summer generations differ from the hibernating forms in being $\frac{1}{3}$ smaller, in their brighter coloring and in the markings of the front wings. The two forms are so unlike that they were formerly considered as distinct species.

Methods of control.

Pear orchards subject to attack should be kept clean of trash under which the hibernating psyllas might find shelter. During the fall or winter the rough bark should be scraped from the trunks and larger branches to render them less attractive as hibernating quarters, and to make it easier to reach the insects by spraying during the dormant season. The scraping can be done with a dull hoe, taking care not to injure the living wood. This work can be done to best advantage during damp weather, when the rough bark comes off the easiest. When the psyllas are abundant great numbers are scraped off with the bark; in such case it might pay to collect and burn the scrapings. During warm days in November and December, or in March and early April, many of the hibernating adults crawl from their hiding places in crevices of the bark. At such times they are sluggish in their movements and do not readily take flight. A large proportion of them can be destroyed by a thorough spraying of the trees with "Black Leaf 40" tobacco extract, 1 pint in 100 gallons of water, adding three to five pounds of soap. The application should be made on days when there is no danger of the liquid freezing on the trees. It is important to spray both sides of a tree before proceeding to the next because the flies often dodge to the opposite side of a branch, and thus escape being hit by the spray. Effective work against the hibernating flies can also be done by using kerosene emulsion diluted with 10 parts of water, miscible oil diluted with 12 or 15 parts of water, or whale-oil soap 1 pound in 4 to 6 gallons of water. These mixtures are, on the whole, not so satisfactory as the tobacco extract from the standpoint of safety and efficiency.

Many of the eggs and newly hatched nymphs can be destroyed by spraying with lime-sulfur at the strength used for scale, making the application when the blossom cluster-buds are beginning to separate at the tips.

By the time the petals have fallen, nearly all of the eggs have hatched and the young nymphs are clustered at the base of the leaf petioles and the fruit stems, where they may be easily killed by thorough spraying with kerosene emulsion, diluted with 10 parts of water, whale-oil soap, 1 pound in 4 to 6 gallons of water, or " Black Leaf 40 " tobacco extract, $\frac{3}{4}$ pint in 100 gallons of water, adding 3 to 5 pounds of soap. As the nymphs grow older they become more difficult to kill and the expanding foliage gives them more protection from the spray. To get the best results the first application should be made just after the petals fall and should be repeated in three or four days. By thoroughly killing off the first brood at this time complete protection of the crop for the entire season can be obtained. It has never been found practicable to kill the summer adults; they are very active and take flight at the slightest alarm.

References

Cornell Agr. Exp. Sta. Bull. 44. 1892.
Cornell Agr. Exp. Sta. Bull. 108, pp. 69–81. 1896.
N. Y. (Geneva) Agr. Exp. Sta. Circ. 20. 1913.

False Tarnished Plant-bug

Lygus invitus Say

This close relative of the tarnished plant-bug has for many years caused considerable loss to the pear growers of western New York. In certain orchards half of the crop has been rendered unsalable by the punctures of this obscure insect. Pears that have been injured in this way are knotty, deformed and gritty in texture (Fig. 203).

The adult is about ¼ inch in length and light brownish in color. The winter is passed in the egg stage. The eggs are doubtless inserted in the bark of the smaller branches; they hatch during the blossoming period. The pale-colored nymphs at first feed on the tender opening leaves, but attack the fruit as soon as it sets. After the first molt they take on a greenish color, which makes them very inconspicuous as they rest on the leaves or young fruits. They pass through five immature stages, becoming mature about the middle of June. Most of the injury is done towards the last of May while the fruits are still quite small. A single nymph may visit many pears, puncturing each one several times. The tissue surrounding the puncture hardens, becomes gritty, and the growth of the fruit at that point is retarded, causing a depression in the surface. The location of the puncture is usually indicated by a break in the skin, from which protrudes a small yellowish granular mass. Badly punctured fruits are often considerably undersized.

Fig. 203. — Mature pear showing the result of injury by the false tarnished plant-bug.

Control. — The newly hatched nymphs are tender, delicate creatures and may be easily killed by a thorough application of " Black Leaf 40 " tobacco extract, ¾ pint in 100 gallons of water, adding three or four pounds of soap to make the liquid stick and spread better. This application should be made about the time the petals are falling. In some cases it may be necessary to repeat the treatment a few days later.

REFERENCE

N. Y. (Geneva) Agr. Exp. Sta. Bull. 368. 1913.

The Pear Thrips

Euthrips pyri Daniel

In the San Francisco Bay region of California, since 1904, pears, prunes, apricots, peaches and almonds have suffered greatly from the attacks of this minute, fringe-winged insect. More recently this pest has appeared in injurious numbers both in New York state and in England. The dark brown adults, only about $\frac{1}{20}$ inch in length, emerge from the ground and attack the bursting buds of the fruit trees in late February and early March in California. They work their way into the opening buds and feed on the tenderest parts of the unopened leaves and blossoms. In feeding they pierce and rasp away the epidermis with their mouth parts and then suck out the sap. The female inserts her minute, whitish, bean-shaped eggs principally in the stems of the young fruit and leaves. The eggs hatch in about four days. The young nymphs are white with red eyes, closely resembling the adults in form, except for the lack of wings, and have similar feeding habits. They attain their growth in two or three weeks and fall to the ground, where, at a depth of several inches, they form a small earthen cell within which they hibernate. Most of the cells are found in the first 3 or 4 inches of firm soil; none are made in the loose surface layer. During the winter months the nymphs change to adults within the cells, but the latter do not emerge till the last of February in California.

The injury is greater on those fruits which, like the pear and prune, bear the blossoms in clusters all developed from one bud, than on the peach and apricot, where only one blossom comes from a single bud, because in the former case the thrips have more chance to work before the fruit sets. Pears and prunes suffer most severely. When abundant a large number of thrips will enter a single bud, stunt the leaves, blast the blossoms and prevent the setting of the fruit. On pears most of the injury is done in the bud itself; on prunes and peaches the nymphs feed on the skin of the young fruits beneath the drying calices, causing scabby or silvered areas. On both prunes and cherries much injury is caused by the egg-laying punctures in the fruit stems; the fruits turn yellow and drop. On all fruits the injury to the foliage through a number of successive years tends to weaken the trees and render them subject to disease.

Remedial measures.

The pear thrips may be satisfactorily controlled by proper cultivation and spraying. During October, November and December the ground should be plowed to a depth of 7 to 10 inches, harrowed, cross-plowed and again harrowed. This is intended to break open the cells and kill the tender last stage nymphs. This method has given good results in California prune orchards, but is less efficient in pear orchards.

Two sprayings should be made, the first just as the first buds begin to open and the second just after the petals fall, using $\frac{3}{4}$ pint "Black Leaf 40" tobacco extract in 100 gallons of water, adding 5 pounds of soap. In the first spraying the object is to force the liquid into the opening buds where it will kill the thrips by contact. To do this it is necessary to use high pressure, at least 150 pounds, and a rather coarse nozzle, spraying down directly into the bud tips. In cases of severe infestation it is sometimes advisable to make an additional application directly after the first. Experiments in California have shown that the thrips can be satisfactorily controlled by thoroughly spraying the trees with a thick whitewash just as the buds are opening. The whitewash is made by slaking 80 pounds of quicklime for each 100 gallons of the wash. It should be strained before using.

REFERENCES

U. S. Bur. Ent. Bull. 68, Pt. I. 1909.
U. S. Bur. Ent. Bull. 80, Pt. IV. 1909.
N. Y. (Geneva) Agr. Exp. Sta. Bull. 343. **1912.**
Cal. Agr. Exp. Sta. Bull. 228. 1912.

The Pear Midge

Contarinia pyrivora Riley

This European enemy of the pear was introduced into Connecticut about 1877 and has spread into New York and New Jersey. It attacks all varieties of pears, but has a decided preference for the Lawrence. Its spread has been rather slow and it has not become as serious a check to the pear industry as was anticipated.

The adult insect is a small midge, about $\frac{1}{10}$ inch in length, closely resembling a mosquito. They appear about the time the blossoms begin to show color, and the female deposits, by means of her long flexible ovipositor, a mass of 12 to 45 minute, yellowish-white, elongate eggs in the interior of the unopened blossom. The eggs hatch in 4 days or more, depending on the temperature; and the minute, whitish larvæ work their way down into the ovary, destroy the core and

Fig. 204.—A young pear cut open to show larvæ of the pear midge. Enlarged.

hollow out a large irregular cavity, which frequently occupies nearly the whole interior of the young fruit (Fig. 204). Infested fruits at first become abnormally enlarged and later stunted and deformed. The larvæ become full grown early in June and escape from the fruit usually through cracks which occur after heavy rains (Fig. 205). In dry seasons the fruit may fall before cracking open; in such cases the

Q

larvæ bore their way out through decayed spots. They burrow into the soil an inch or two, and after a variable time spin delicate cocoons within which the insect hibernates. Most of the larvæ spin cocoons within a month after entering the ground, but some are found naked as late as October. As a general thing the winter is passed in the pupal stage, but sometimes a large percentage of the larvæ may not pupate until spring, about two weeks before the flies emerge. When about to transform the pupa leaves the cocoon and works itself to the surface of the ground, where the fly is set free. There is only one generation a year, although it has been observed in France that some of the first larvæ to pupate transform to flies during July of the same year. As there are no pear blossoms available at that time and as they have no other food-plant, these precocious individuals necessarily perish.

FIG. 205. — Young pears infested by the pear midge, cracked open to permit the escape of the larvæ.

Remedial measures.

So far no method of spraying has been devised to prevent the laying of the eggs or the development of the larvæ in the fruit. The infested fruits are easily distinguished by their size and shape, and where only a few trees are affected it would pay to collect and burn them before the larvæ have emerged, that is before May 15. In larger orchards this would be too expensive to be practicable. Experiments in New Jersey have shown the great value of kainit in destroying the larvæ in the ground. It should be applied during the latter half of June, at the rate of 1000 to 2000 pounds to the acre. While it has been successfully used on the sandy soils of New Jersey, the experience of other growers indicate that such heavy applications of kainit on clay soils is very liable to injure the trees. In such cases recourse must be had to repeated, thorough, shallow cultivation during June and July. By this means a large number of the larvæ and pupæ will be destroyed.

REFERENCES

Riley, Rept. U. S. Com. Agr., pp. 283–289. 1885.
N. J. Agr. Exp. Sta. Bull. 99. 1894.
Marchal, Ann. Soc. Ent. Fr., pp. 5–27. 1907.

The Pear-leaf Blister-mite

Eriophyes pyri Pagenstecher

The leaves of pears and apples are often disfigured by reddish or greenish-yellow blisters which later in the season turn brown (Fig. 206). These blisters are caused by colonies of minute, whitish, elongate, four-legged mites, $\frac{1}{125}$ inch in length, living within the tissue of the leaf (Fig. 207). These creatures are not insects, but belong to the class of animals known as Arachnida, to which belong spiders and scorpions. While originally a native of Europe, it has now become widely distributed and occurs wherever the pear is cultivated. About 1902 it suddenly became an important apple pest in the Eastern states and Canada. The cause of this remarkable change of habit is unknown.

The adult mites pass the winter snugly hidden away beneath the second or third bud scales. With the bursting of the buds they migrate to the tender leaves and burrow beneath the epidermis of the under surface. The irritation thus caused produces a thickening of the leaf tissue and results in the formation of a blister-like gall. The minute, whitish eggs are deposited within the gall, and the young remain there until mature. They then leave the gall through a minute opening on the under side, migrate to new leaves and there start new blisters. Reproduction is continuous and new galls are formed throughout the growing season.

Fig. 206. — Pear leaves infested with the pear-leaf blister-mite.

On pear the blisters at first appear as small greenish pimples which soon take on a reddish color and later turn brownish.

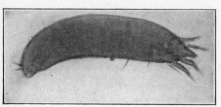

Fig. 207. — Pear-leaf blister-mite, greatly enlarged.

On apple the first stages are pale yellowish and the blisters never become as red as on pear. They are $\frac{1}{8}$ inch or less in diameter, and when abundant coalesce, producing large dead areas in the leaves. Badly infested leaves are liable to turn yellow and drop. The loss of foliage weakens the tree, interferes with the maturing of the fruit and the formation of fruit buds.

Pear trees in the nursery are sometimes badly stunted in this way. Sometimes the mites attack the very young fruits (Fig. 208) and fruit stems, causing small pimples, but the injury is usually outgrown and little or no loss results.

Remedial treatment.

The leaf blister-mite is not a difficult pest to control. Lime-sulfur, miscible oils and homemade oil emulsions have all given excellent results; but lime-sulfur has, on the whole, proved most satisfactory and is now extensively used by commercial orchardists. The insecticide is intended to destroy the adults hibernating under the bud scales and can be applied either in the fall after the leaves have

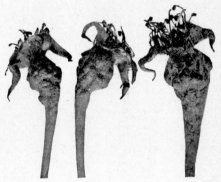

FIG. 208. — Young fruits deformed by the blister-mite.

fallen or in the spring any time before the tips of the leaves begin to show. When used for blister mite alone the lime-sulfur may be applied somewhat weaker than for the San José scale; a dilution of 1 to 10 when the concentrated solution tests 31° Beaumé is sufficiently strong. As a rule it is not necessary to spray every year for the control of blister-mite; one treatment usually so reduces the number of mites that they do not again become abundant enough to cause serious injury for several years.

REFERENCES

N. Y. (Geneva) Agr. Exp. Sta. Bull. 283.   1906.
N. Y. (Geneva) Agr. Exp. Sta. Bull. 306.   1908.

The Sinuate Pear Borer

Agrilus sinuatus Olivier

When this European enemy of the pear was first discovered in New Jersey, in 1894, it was greatly feared that it would become widely distributed and as seriously interfere with the pear in-

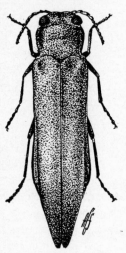

Fig. 209. — The sinuate pear-borer beetle (× 7½).

Fig. 210. — Pear branch infested with the sinuate borer.

dustry in this country as it does in Germany and France. Fortunately such has not been the case. As far as known it is still confined to that state and New York.

The slender, shining, bronze-brown beetles (Fig. 209) about ⅓ inch in length, emerge the last of May and during June. They are found on bright sunny days on the bark of the trunk and branches, where the female deposits her eggs in crevices and under flakes of bark. The eggs hatch in early July and the slender whitish grubs eat out narrow winding burrows in the sapwood. The partly grown larva rests in its burrow during the first winter and the next spring continues its destructive work. The burrows are now much larger and more winding; they frequently intersect, cut off the supply of sap and kill the branch or tree. The course of the burrows is now indicated exteriorly by the discolored and abnormal bark above them (Fig. 210). In smooth-barked trees these winding trails become very conspicuous. In September of the second year the larva, which is now about $1\frac{1}{2}$ inches in length, burrows into the solid wood and there at the depth of about $\frac{1}{4}$ inch constructs a pupal chamber which it connects with the bark by an exit hole. It then plugs both ends of the chamber with sawdust, becomes shorter and thicker, and the following April transforms to a pupa. The beetles emerge about a month later.

Trees of all sizes from nursery stock up are subject to attack. Smaller trees are frequently completely girdled by the intersecting burrows and killed outright. Larger trees are so weakened by the presence of the borers that they soon present a sickly appearance, lose their leaves and finally die. The Kieffer is less subject to injury than other varieties because of its greater vitality, whereby it is able to fill up the burrows with new tissue.

Treatment.

Infested nursery stock and all other trees too badly infested to be of value should be removed and burned. In many cases the pupal chamber can be located by a discolored area in the bark and the insect dug out with a knife. Attempts to keep the beetles from emerging by coating the bark with a viscid wash have not been successful, but it is probable that deterrent washes, as suggested for the Round-headed Apple Tree Borer, would be of some value in preventing oviposition. Several applications should be made during the last part of May and early June, taking care to cover the larger branches as well as the trunk.　Trees that have been kept in strong, healthy condition by proper care and the use of appropriate fertilizers are better able to withstand borer attack.　As this pest is readily distributed in nursery stock, one should be careful in setting a new orchard to have all young trees carefully inspected.

REFERENCE

N. J. Agr. Exp. Sta. 15th Ann. Rept. (1894), pp. 550–551.　1895.

The pear borer (*Sesia pyri* Harris) occurs throughout the Eastern states and Canada, ranging southward to Florida and Texas.　It is a near relative of the peach tree borer.　The adult is a clear-winged moth with an expanse of $\frac{3}{4}$ inch, bluish-black in color, and having the abdomen marked with three yellow bands; in the female the anal tuft is also yellow.　The larvæ, which closely resemble those of the peach tree borer, burrow in or just below the bark on the trunk and larger branches of the pear and apple, but do not penetrate the sapwood as is usual in that species.　It, therefore, rarely causes serious injury.

PEAR-BLIGHT BEETLE

Xyleborus dispar Fabricius

Sometimes in June the tips of pear and apple branches suddenly die back as the result of the work of a small Scolytid beetle.　The injury is often mistaken for the bacterial disease, pear blight, whence the common name.　These beetles belong to that highly interesting group of wood borers known as ambrosia beetles from their habit of feeding, both as larvæ and adults, on a peculiar fungus propagated in their burrows and known as ambrosia.　They are also sometimes known as wood stainers from the fact that the growth of the fungus blackens the surrounding wood.

This species occurs in Europe, Northern Asia, the northern United States and Canada; it infests among other forest trees the hemlock, beech, birch, red oak, and among fruit trees the pear, plum and apple, preferring the latter.

The female beetle is dark brown, about $\frac{1}{8}$ inch in length, cylindrical and has the nearly globular head drawn under the thorax so as to be invisible from above. The males are only a little more than half as long and have the back very convex longitudinally. On small branches the female usually starts her burrow just below a bud scar and, after passing through the sapwood tunnels around the pith, keeping in the hard wood, constructs a number of side burrows running lengthwise of the branch. In larger branches the galleries are straighter. The eggs are laid loosely in the burrows in June in Nova Scotia and in May in West Virginia, and the grubs feed on the fungus growing on the walls of the chamber. The larvæ pupate in the galleries and the beetles escape through the entrance hole made by the parent beetle.

Treatment.

So far this beetle has been troublesome only in restricted localities and is not likely to become a serious pest. In Nova Scotia a wash made of 3 gallons of water, 1 gallon soft soap and $\frac{1}{2}$ pint of carbolic acid has given good results when applied in June. In using this wash the aim is to have the liquid soak into the burrow, kill the food fungus and thus indirectly destroy the beetles. Two or three applications at intervals of a few days are advised. Badly infested trees or branches should be burned before the beetles have a chance to emerge.

REFERENCES

Peck, Mass. Agr. Jour. IV, pp. 205–207. 1817.
Hubbard, Bur. Ent. Bull. 7, pp. 9–30. 1897. General account of the ambrosia beetles.
Fletcher, Rept. Ent. Bot. for 1904, pp. 240–241. 1905.
Swaine, Rept. Ent. Soc. Ont. for 1909, pp. 58–63. 1910.
Schneider-Orelli, Centlb. Bakt. Parasitenk. Infektk., 2 Abt., XXXVIII, pp. 25–110. 1913. Extensive account.

The Howard Scale

Aspidiotus howardi Cockerell

This scale insect was first discovered in Colorado in 1894, and now ranks as a serious pest in pear, plum and prune orchards in that state. It also attacks apple, peach, wild plum, white ash and maple trees; and has been found in New Mexico. Besides infesting the bark, especially on the twigs, it often occurs on the fruit, causing a peculiar pitting of the surface on pears, and, like the San José scale, reddish discolorations of the skin around the scales. Trees are sometimes incrusted and killed by this scale. It is closely allied to the San José, the Putnam's, the cherry and the European fruit scales, being distinguished with certainty by microscopic characters only. Mature female scales are circular, about the same size as the San José scale, but of a paler grayish color with the dull orange exuvial spot on one side of the center. In Colorado there are three and possibly four overlapping generations of this scale annually, the winter being passed in the partly grown condition. Males emerge early in April and six or eight weeks later crawling young appear. The species is apparently both oviparous and ovoviviparous, usually eggs, but sometimes only minute, living young, being found under the mother scales.

This Howard scale spreads slowly and is attacked by a little parasite, *Prospaltella aurantii*, and the twice-stabbed lady-bird beetle as well as small spiders destroy many. The lime-sulfur wash applied late in spring before the buds open has been found to effectively control the insect in Colorado.

The European Pear Scale

Epidiaspis piricola Del Guercio

This small, circular European scale insect has been sent to the United States several times on nursery stock, but seems to have established itself in injurious numbers only in California, although recorded also from several Eastern states. It has attacked pear and prune trees principally in this country, but the apple, plum, peach and currant are among its food-plants. Specimens we saw on pear seedlings from France were in little pits or depressions in the bark, and others state they occur under the edges of rough bark or moss. Apparently it spreads and breeds slowly, so it will doubtless never be a serious menace to American orchards. The winter applications recommended for the San José scale will probably control it if thoroughly applied.

OTHER PEAR INSECTS

CODLIN-MOTH : *apple*, p. 10.
GREEN FRUIT-WORMS : *apple*, p. 39.
BUD-MOTH : *apple*, p. 42.
CIGAR-CASE-BEARER : *apple*, p. 47.
CLICK-BEETLES : *apple*, p. 55.
FRUIT-TREE LEAF-ROLLER : *apple*, p. 62.
OBLIQUE-BANDED LEAF-ROLLER : *apple*, p. 65.
LEAF-CRUMPLER : *apple*, p. 68.
UNSPOTTED TENTIFORM LEAF-MINER : *apple*, p. 73.
SERPENTINE LEAF-MINER : *apple*, p. 74.
RESPLENDENT SHIELD-BEARER : *apple*, p. 75.
WHITE-MARKED TUSSOCK-MOTH : *apple*, p. 100.
ORIENTAL MOTH : *apple*, p. 106.
FALL WEBWORM : *apple*, p. 107.
APPLE-TREE TENT-CATERPILLAR : *apple*, p. 112.
FOREST TENT-CATERPILLAR : *apple*, p. 119.
YELLOW-NECKED APPLE CATERPILLAR : *apple*, p. 123.
RED-HUMPED APPLE CATERPILLAR : *apple*, p. 125.
CLIMBING CUTWORMS : *apple*, p.138.
APPLE LEAF-APHIS : *apple*, p. 147.
ROSY APPLE APHIS : *apple*, p. 149.
APPLE BUD-APHIS : *apple*, p. 151.
WOOLLY APHIS : *apple*, p. 153.
SAN JOSÉ SCALE : *apple*, p. 162.
OYSTER-SHELL SCALE : *apple*, p. 171.
SCURFY SCALE : *apple*, p. 176.

QUINCE INSECTS

Nearly all the insects attacking the quince have been discussed under the apple. One of these, the round-headed apple-tree borer, is even more destructive to the quince than to the apple. The only insect which is distinctly a quince pest is the quince curculio treated below.

The Quince Curculio

Conotrachelus cratægi Walsh

This is by all odds the most destructive insect with which the quince grower has to contend. In unprotected orchards of western New York often over 90 per cent of the crop is either infested by the grubs or rendered gnarled and knotty by the punctures of the beetles (Fig. 218).

Fig. 211. — The quince curculio, side view.

Fig. 212. — The quince curculio, dorsal view (× 3).

The quince curculio is a brownish-gray, broad-shouldered snout beetle, about $\frac{1}{4}$ inch in length (Figs. 211 and 212); the wing cases are strongly ribbed lengthwise by sharp ridges, and there are two rows of deep punctures in each interval; there are no humps as in the case of the plum and apple curculios. The time of the first appearance of the weevils on the trees varies greatly with the season. In 1896 they appeared during the last week in May, while in 1897 they did not begin work until about two months later or the last of July.

Fig. 213. — Young quince, showing the manner
of feeding of the curculios.

In feeding the beetle cuts a small opening in the skin of the fruit with the jaws at the tip of the snout and then eats out a cavity in the pulp (Fig. 213). The small opening to the cavity is not readily seen, as it is usually hidden by the thick fuzzy coating of the fruit. These early punctures are the cause of most

Fig. 214.—Section through the egg-pit, showing the egg in position, much enlarged.

Fig. 215. — Larva of the quince curculio, enlarged.

of the knotty and deformed fruits. The cavity becomes filled with hardened, gritty tissue, growth is stopped at that point, and the scar of the puncture comes to occupy the bottom of a deep depression in the surface of the fruit.

Fig. 216. — Pupa of the quince curculio (× 4).

The female deposits her minute, whitish, oval eggs singly in cavities (Fig. 214) indistinguishable from those excavated for food only. The eggs hatch in from seven to ten days, and the grub burrows through the flesh, seldom reaching the core. Only a small proportion of the infested quinces fall to the ground, and most of the grubs develop in fruit still hanging on the tree. In about 30 days the full-grown, flesh-colored, footless, maggot-like grub (Fig. 215) leaves the fruit and burrows two or three inches into the soil. There it remains in the grub stage, in a small earthen cell until the next spring, when it transforms to the tender and helpless pupa (Fig. 216). From 10 to 20 days are spent in this stage before the change to the

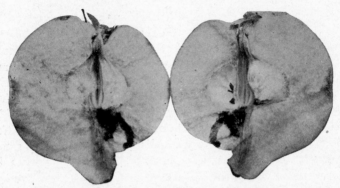

Fig. 217. — A "wormy" quince cut open.

beetle takes place. The beetle is at first white and soft and remains within the cell for 10 days or more before it is strong enough to work its way to the surface. There is only one generation a year. In normal seasons egg-laying probably begins about the middle of July and continues for a considerable period. The number of eggs laid by one female has not been determined, but probably

Fig. 218. — A "knotty" quince, the result of curculio work.

exceeds a hundred. The grubs begin to emerge in August, but many are still in the fruit at picking time.

Treatment.

The quince curculio is a difficult insect to control. It passes the winter as a grub in the soil, and therefore the destruction of hibernating quarters as advised for the plum and apple curculios would be of no avail. Clean cultivation has not given the expected results in the destruction of the grubs, because they are more active than those of the other species and are able to burrow back into the soil and reconstruct their cells after having been disturbed. Shallow repeated cultivation at the time the helpless pupæ are in their cells would probably kill many, but enough escape to make the operation of doubtful value. The destruction of windfalls would reach only a small proportion of the grubs, since most of the infested quinces do not drop, but remain on the tree until after the grubs have become full-grown and emerged. Some commercial growers have had good success in reducing the amount of infestation by picking off and destroying all infested fruit about a month before picking time thus leaving on the trees only first and second class quinces.

Catching the beetles by jarring them on to sheets or curculio catchers is a rather expensive and laborious operation, but has been profitably practiced by extensive growers in western New York. The curculio catcher as used in New York is a large funnel-shaped frame covered with canvas and mounted on a two-wheeled wheelbarrow. In front is a narrow opening reaching to the center, designed to admit the trunk of the tree. The machine is placed in position, and the tree is jarred with a padded mallet having a long handle. The beetles feign death, fall on to the sheet and are caught in a box or can placed under the center of the funnel. From time to time the beetles are removed and killed. The trees should be gone over every day or two while the beetles are present. To determine when they appear, jar a tree or two daily, beginning about the last of May. It is a more difficult matter to jar low-headed, wide-spreading quince trees than the more erect plum trees, and this method of control is now little used in commercial orchards.

The beetles are hard to kill by spraying with an arsenical, because they eat but very little of the skin of the fruit in making their punctures and because of the dense fuzzy covering of the fruit on which the poison is deposited. The use of Paris green or London purple has not given satisfactory results, but there is considerable evidence to show that spraying with the more adhesive arsenate of lead is of considerable value in reducing the number of punctures. It should be used at the rate of 5 or 6 pounds to 100 gallons of water or combined with the proper fungicide. Two applications should be made, the first when the beetles first appear and the second about a week later.

REFERENCE

Cornell Agr. Exp. Sta. Bull. 148. 1898.

OTHER QUINCE INSECTS

CODLIN-MOTH : *apple*, p. 10.
GREEN FRUIT-WORMS : *apple*, p. 39.
BUD-MOTH : *apple*, p. 42.
FRUIT-TREE LEAF-ROLLER : *apple*, p. 62.
APPLE LEAF-SKELETONIZER : *apple*, p. 67.
LEAF-CRUMPLER : *apple*, p. 68.
RESPLENDENT SHIELD-BEARER : *apple*, p. 75.
WHITE-MARKED TUSSOCK-MOTH : *apple*, p. 100.
YELLOW-NECKED APPLE CATERPILLAR : *apple*, p. 123.
APPLE LEAF-APHIS : *apple*, p. 147.
APPLE BUD-APHIS : *apple*, p. 151.
WOOLLY APHIS : *apple*, p. 153.
BUFFALO TREE-HOPPER : *apple*, p. 160.
SAN JOSÉ SCALE : *apple*, p. 162.
OYSTER-SHELL SCALE : *apple*, p. 171.
SCURFY SCALE : *apple*, p. 176.
ROUND-HEADED APPLE-TREE BORER : *apple*, p. 185.
FLAT-HEADED APPLE-TREE BORER : *apple*, p. 194.
TWIG-PRUNER : *apple*, p. 200.

R

CHAPTER VII

PLUM INSECTS

Plums and prunes are subject to attack by the same set of insects. The most troublesome enemies of these fruits are the plum curculio, the San José scale (p. 162) and the European fruit lecanium. The curculio is primarily an enemy of stone fruits, but also attacks the apple, pear and quince. The fruit lecanium, while a very general feeder, seems to be most destructive to the plum. With the possible exception of curculio injury in some locations, it is not difficult to effectively protect a plum crop from insect attack by following a proper system of spraying.

The Plum Curculio

Conotrachelus nenuphar Herbst

This native American snout-beetle or weevil whose original food was the fruit of the wild plum and hawthorn is generally distributed over the Eastern states and Canada, east of the Rocky Mountains, and is a serious pest east of the 100th meridian and occurs southward to Texas and Florida. It attacks plums, prunes, cherries, peaches, nectarines and apricots among stone fruits, and is also a serious enemy of apples and will attack pears and quinces. We have also reared it from gooseberries in New York.

It is by far the most destructive insect with which the grower of stone fruits has to contend, particularly in the South, where it is especially injurious to peaches, often destroying the entire crop in unprotected orchards. Quaintance estimates the

243

annual loss occasioned by this insect at over $8\frac{1}{2}$ million dollars. Although the plum curculio does not breed as freely in apples as in stone fruits, still these are often seriously injured by its feeding and egg-laying punctures. A large proportion of those stung fall early in the season, while those that remain on the trees are usually rendered knotty and unmarketable. It has

FIG. 219. — The plum curculio, side and dorsal view ($\times$ 5).

FIG. 220. — Egg of the plum curculio in position, the skin of the fruit removed. Section through an egg-scar of the plum curculio showing the egg position. Enlarged.

been recognized as a serious pest for over a century, and fully satisfactory means of control have not yet been devised. Control is particularly difficult because the greater part of the injury is caused by the adult, a long-lived, hard shelled beetle very difficult to poison because it feeds principally on the pulp of the fruit obtained through a small puncture in the skin.

The insect passes the winter in the beetle stage, hidden away under leaves or other trash. Stone walls or hedges and adjoining wood lots furnish ideal hibernating quarters, as is shown by the greater injury to that part of orchards lying nearest to such retreats. In the spring about the time the buds open the beetles desert their winter quarters and appear on the trees. They are small, rough snout-beetles, about $\frac{1}{5}$ inch in length, mottled with black, gray and brownish, and there is a black shining hump on the middle of each wing cover (Fig. 219).

The sharp-biting jaws are located at the tip of the snout, which hangs down something like the trunk of an elephant. The beetles attack the fruit as soon as it is set. Two kinds of punctures are made: those for feeding only and those for the reception of the egg. In feeding the beetle cuts a small, round

FIG. 221. — Plum curculio egg-crescents in young plums.

opening through the skin and then eats out a cavity in the pulp about $\frac{1}{8}$ inch in depth, or as deep as it can reach with the tip of the snout. In egg-laying the female first makes a cut through the skin of the fruit and runs her snout obliquely into the flesh just under the skin and gouges out a cavity large enough to receive her egg. Then turning around, a minute, white egg is dropped into the hole, and reversing her position she pushes it into the cavity with her snout. Just in front of the hole she now cuts a crescent-shaped slit which she extends obliquely underneath the egg cavity so as to leave the egg in a flap of the flesh (Fig. 220). Each female may lay from 100 to over

500 eggs, and the egg-laying period may extend over nearly the whole season, although the great majority are laid during the first month after the fruit sets. Fruit marked by these characteristic egg punctures is shown in Figures 221 and 222.

The eggs hatch in 3 to 7 days, and the white or yellowish grub burrows through the flesh, and in the case of stone fruits comes to lie next the pit. Infested fruits, most varieties of cherries excepted, usually fall to the ground before the grubs

FIG. 222. — Plum curculio egg-crescents in young apples.

mature. In the case of apples many of the young grubs are killed in their burrows by the pressure of the growing fruit cells, for the stopping of growth seems to be necessary for their proper development.

When full-grown, which requires about twenty days from the time the egg is laid, the larva leaves the fruit and burrows a short dis-

FIG. 223. — Feeding punctures of the plum curculio in plums made in August.

tance, not over an inch or two, into the ground, and there constructs an earthen cell within which it transforms to a pupa.

About 28 days, on an average, after entering the ground the beetles emerge, but not all this time is passed as a pupa; 12 to 16 days are spent in the soil before pupation, and after the transformation to the adult it takes several days for the beetle to become hardened enough to work its way to the surface. These beetles of the new generation do not as a rule lay eggs the same season, but after feeding greedily on the fruit (Fig. 223) for some time seek hibernating quarters on the approach of cold weather. In the North, the damage done by these new beetles may be very great, especially on plums and apples, but in the South, where early maturing varieties are grown, it is of less importance.

Methods of control.

Curculios thrive in neglected, over-grown, unpruned and uncultivated orchards surrounded by stone walls or neglected hedges and situated near pieces of woodland. The first step towards their control consists in correcting as far as possible all these conditions at variance with the best horticultural practice. Hibernating shelter should be reduced to a minimum by the removal of

Fig. 224. — Plum exuding gum as a result of curculio injury.

all stone walls and stone piles, by the cleaning up of overgrown hedges and fences and by the destruction of all trash under which the beetles might find shelter. The trees should be properly pruned to admit the sun, for curculios are shy creatures, preferring the deep shade for their work, and furthermore it has been shown in Illinois that direct sunlight striking the fallen fruit is soon fatal to the grubs within. Where the trees are large and the ground is kept smooth and free from weeds, it is worth while to rake the fallen fruits out into the open where the sun can strike them. In doing this the small early drops should not be neglected, for a large percentage of the grubs develop in fruit not much larger than peas.

Frequent and thorough, though shallow, cultivation during the period that the tender and helpless pupæ are in the ground is of great value. If their cells are broken open, the pupæ are either killed at once or soon fall a prey to their natural enemies, principally ants and beetle larvæ. To obtain the best results, cultivation should be continued for about a month or six weeks after the grubs begin to go into the ground. In the North this will be from about July 10 to August 10, while in the South it is somewhat earlier. As most of the pupal cells are within an inch or two of the surface, the cultivation may be quite shallow, but it should be thorough, and care should be taken to stir the ground close under the trees where most of the pupæ lie. Most growers in the North do not favor cultivation of orchards after August 1, but where the curculio is abundant, the cultivation should be continued as late as possible without interfering with the maturing of the wood and fruit.

Jarring.

For many years this was the most satisfactory method of fighting the curculio on plums and peaches, but is now rarely practiced in commercial orchards. If the tree is suddenly jarred with a padded mallet, the beetles will loosen their hold, contract the legs and fall to the ground, feigning death for a considerable time. Where only a few trees are to be treated, the beetles may be caught on a large sheet spread beneath the tree. Where the ground is fairly smooth and where the trees are headed high enough, the work may be done more quickly and easily by using a wheeled curculio catcher, as described under quince curculio, p. 240. These machines have been used for many years by the plum growers of western New York. In the extensive Georgia peach orchards jarring has been practiced on a large scale. There the curculios are caught on sheets stretched on light frames, 12 feet long by 6 feet wide. Two frames are held under the tree by four persons, while the fifth jars the tree with a padded mallet. When the end of the row is reached, the curculios are picked out for destruction, while the beneficial lady-bird beetles are allowed to escape.

To be most effective, the jarring should be done very early in the morning, for then the beetles are less active, they fall more readily and are less liable to escape from the sheets. The trees should be gone over every morning for four or five weeks, or until no more beetles are captured. Unless cheap labor can be obtained readily, jarring is too expensive for large orchards and in such cases has been generally supplanted by improved methods of spraying.

Spraying.

Spraying with arsenate of lead, either alone or combined with a fungicide, has now come to be the favorite method of fighting the curculio on most crops. It has replaced Paris green for this purpose because it adheres better, is less liable to injure the foliage and can be combined with lime-sulfur when used as a fungicide. In general, the foliage of stone fruits is very susceptible to injury from soluble arsenic, even when present in small amounts, and repeated heavy applications of arsenate of lead are attended with some risk. Nevertheless, the experience of the past few years has shown that this danger may be avoided by following the system of spraying suggested below.

Plums.

Although the control of the curculio on plums by spraying is more difficult than on most other fruits, still many growers believe that the results justify the practice. Two applications should be made, the first soon after the petals fall and the second a week or ten days later, using arsenate of lead, $2\frac{1}{2}$ pounds in 50 gallons of Bordeaux mixture (2–3–50) or self-boiled lime-sulfur.

Cherries.

In general, spraying for curculio on cherries has given better results than on plum. The applications should be made as indicated above.

Peaches.

On peaches the direct injury caused by the curculios is greatly augmented by the brown rot which is distributed by the beetles and which gains entrance to the fruit through their punctures. As far as spraying is concerned, the two problems are considered as one, and the following system has been devised for the control of both, based on extensive experiments in Georgia and Missouri.

The first application should be made just as the calyxes (shucks) are shedding, using 2 pounds arsenate of lead to 50 gallons of water, to which is added the milk of lime made from slaking 2 pounds of stone lime. As this is too early for brown rot, the fungicide is not used.

The second application is made about three weeks later, using 2 pounds of arsenate of lead in 50 gallons of the self-boiled lime-sulfur (8–8–50).

A third application should be made about one month before the ripening of the fruit, using self-boiled lime-sulfur only. The poison is omitted, because experience has shown that it is very unsafe to spray peaches more than twice with arsenate of lead.

Apples.

In spraying apples there is practically no danger of injury to the fruit or foliage by repeated applications of arsenate of lead. The two sprayings usually given for the codlin-moth, just after the petals fall and three weeks later, are of considerable value in controlling the curculio, but where the infestation is severe, additional applications will be found necessary. The spray often given for the second brood of the codlin-moth nine or ten weeks after the petals fall will help to control the pest, and in the South it is sometimes advisable to repeat the application two or three weeks later. To get the best results the spraying must be done in a very thorough manner, using a fine nozzle and high pressure in order to keep the fruit evenly coated with the poison. It has been shown by extensive experiments in Illinois that where this plan is followed and other conditions are favorable, the curculio injury may be reduced from 20 to 40 per cent.

In fighting the curculio, reliance should not be placed on any one method of attack. Clean farming to reduce available winter shelter, proper pruning to admit the sun, thorough cultivation at the proper time to destroy the pupæ in the soil, the use of fertilizers to produce strong, healthy trees resistant to spray injury, are all important factors in the fight. In orchards so treated the number of curculios will be reduced to a minimum, and the orchardist will have the best chance to protect his crop by spraying.

REFERENCES

Riley and Howard, Rept. Com. Agr. for 1888, pp. 57–79.
Ill. Agr. Exp. Sta. Bull. 98.   1905.
Quaintance, U. S. Dept. Agr. Yearbook, for 1905, pp. 325–330.   1906.
U. S. Bur. Ent. Circ. 73.   1906.
Mo. State Fruit Exp. Sta. Bull. 21.   1909.
Ga. St. Bd. Ent. Bull. 32.   1910.
U. S. Bur. Plant Ind. Bull. 174.   1910.
U. S. Bur. Ent. Bull. 103.   1912.

THE PLUM GOUGER

Coccotorus scutellaris Le Conte

This native snout-beetle attacks plums, prunes and nectarines throughout the North Central states. It may be distinguished from the other fruit-infesting species by the ochre-yellow head, thorax and legs and the dun-colored wing covers, which are entirely without humps. The insect hibernates in the adult state. The beetles appear on the trees in spring somewhat earlier than the plum curculio, and feed for a time on the buds and leaves. In confinement they have been observed to feed almost exclusively on the ovaries of the buds and blossoms, which they reach by puncturing the calyx.

Soon after the fruit has set the beetles begin to feed upon the pulp through small punctures made in the skin. The female deposits her yellowish-white eggs, about $\frac{1}{35}$ inch in diameter, singly in gourd-shaped cavities gouged out in the fruit with her snout. On hatching, the young grubs burrow directly into the pit and feed on the kernel within until full grown. The larva is then milk-white in color with brownish jaws, and is strongly curved. Before changing to a pupa it eats out an exit hole through the hard shell of the pit, to provide for the escape of the future beetle. The pupal stage is passed within the pit, and the beetles emerge from the fruit during late August and September. They do not feed to any extent the first season, but soon desert the trees and go into winter quarters.

Infested plums do not fall or ripen prematurely. The principal injury is caused by the punctures made by the beetles in feeding and egg-laying; gum exudes from the wounds, and deformed and misshapen fruit results. Sometimes in restricted localities the gouger may be more abundant and cause more injury than the plum curculio, but as a general thing it is not a serious pest.

Treatment.

In general the measures suggested for the control of the plum curculio will apply to this species, but the destruction of fallen fruit would be of no avail, because as a rule plums infested by the gouger do not fall prematurely. Jarring is also less satisfactory because the beetles fall less readily and are more liable to escape by flight. Although the results of definite experiments are not available, it is probable that spraying with arsenate of lead as practiced for the plum curculio would do much to hold this pest in check. The first application should be made just before the blossom buds open.

REFERENCES

Walsh, First Rept. State Ent. Ill., pp. 97–104, 1867 (second edition, 1903).
Riley, 3d Rept. State Ent. Mo., pp. 39–42. 1871.
Iowa Agr. Exp. Sta. Bull. 9. 1890.
Mont. Agr. Exp. Sta. Bull. 62, pp. 211–218. 1905.

THE AMERICAN PLUM BORER

Euzophera semifuneralis Walker

While generally distributed throughout the United States and Canada, this insect has only occasionally become of economic importance. In addition to the plum it has been reported as attacking the pear, apple and mountain ash, and the moth has been reared from larvæ feeding on the black-knot of plum.

The parent insect is a small, obscurely colored grayish moth with an expanse of a little less than an inch. They emerge during late May and early June and are rarely seen, being nocturnal in habit. The eggs are unknown, but are probably deposited in cracks or crevices in the bark of the trunk and larger branches. The larvæ eat out winding burrows next the sapwood, cause large dead areas in the bark and in some cases completely girdle the tree. Their presence is usually indicated by the frass thrown out of the entrance to the burrow. When full-grown the larva is about one inch in length and varies in color from dusky greenish to pinkish or reddish. In Delaware the evidence points to three generations annually, adults of the later broods appearing during the latter part of July and in September, but Dr. S. A. Forbes considered the species single brooded in Illinois. The larvæ of the last brood mature in early November and hibernate in small white silken cocoons under flakes of bark or in the frass thrown out at the opening of the burrow. About May 1 they transform to pupæ, and the moths appear about three weeks later.

Treatment.

As the burrows are indicated by the accumulation of frass, and as the larvæ are always near the surface, it is an easy matter to dig them out with a knife. Scraping away the rough bark in the winter would probably destroy many larvæ in their cocoons. The rubbish should be collected and burned, for if left on the ground many of the larvæ would doubtless survive. Trees kept in a strong, healthy state by proper care are in better condition to sustain attack.

REFERENCES

Forbes, 6th Ann. Rept. State Ent. Ill., pp. 26–29. 1891.
Del. Ag. Exp. Sta. Bull. 53, pp. 9–13. 1901

THE PLUM WEB-SPINNING SAWFLY

Neurotoma inconspicua Norton

The larvæ of this sawfly have the peculiar habit of webbing in the leaves of the plum and cherry, forming unsightly nests somewhat like the cherry tortrix. It has been reported from Manitoba, South Dakota and Massachusetts. The adult is a four-winged fly about $\frac{1}{3}$ inch in length with a black body and rufous legs except the black coxæ and tarsi; the wings are hyaline with a faint fuscous band behind the stigma. They appear just as the leaves are expanding, and the female deposits her smooth, lemon-yellow elongate eggs in two or three rows along the midrib on the underside of the young leaves. The eggs hatch in about eight days, and the young larvæ begin feeding on the leaves under cover of webs which they enlarge to inclose fresh leaves as more food is needed. They become full grown in about a month, by which time the webs frequently cover the entire tree. The larva is then about $\frac{3}{4}$ inch in length, grayish above and yellow or pinkish below, with a yellow head and black thoracic shield and anal segment. When full-grown they find their way to the ground, where at a depth of about six inches they pass the winter in earthen cells. Early the following spring the larva works its way up nearly to the surface and changes to a pupa just below the grass roots. The adults appear about May 1 in Massachusetts and the second week of June in South Dakota. There is only one generation a year.

Treatment.

The larvæ are easily killed by a thorough spraying with arsenate of lead, 5 pounds to 100 gallons of water, applied soon after the eggs have hatched and before the large webs are made.

REFERENCES

S. D. Agr. Exp. Sta. Bull. 48. 1896.

Fletcher, Rept. Ent. for 1896, p. 253.

Fernald, Ent. News XIV, pp. 298–302. 1903.

THE PLUM LEAF-BEETLE

Nodonota tristis Olivier

During June and July the foliage of plum, peach, cherry and chokecherry is occasionally attacked by small, shining, steely-blue flea-beetles, about $\frac{1}{6}$ inch in length. The early stages are unknown, but from what is known of the habits of related species the larvæ probably live on the roots of plants. Eggs laid in confinement are described as being elongate oval, smooth, dirty whitish-gray in color and about $\frac{1}{36}$ inch in length. They were laid in a mass of 36 in the fold of a leaf and hatched in about a week.

The injury may be prevented by spraying with arsenate of lead, 4 pounds to 100 gallons of water, when the beetles are first noticed.

REFERENCE

Chittenden, U. S. Bur. Ent. Bull. 19, pp. 93–95. 1899.

The Hop Plant-louse

Phorodon humuli Schrank

This notorious hop pest usually passes the winter and spring on the plum. It is a native of Europe and was introduced into New York about 1863, but now occurs throughout the country wherever its host plants are grown.

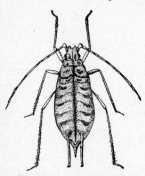

Fig. 225. — The hop plant-louse, a wingless viviparous female of the third generation. Redrawn after Riley. Enlarged.

The black shining oval eggs are deposited around the buds on the terminal twigs of the plum in autumn and hatch soon after the buds open the following spring. The light green lice cluster on the leaves and tips of the tender branches, sometimes severely injuring the crop. Usually, there are about three generations of wingless viviparous females (Fig. 225) on the plum and then winged forms are produced which migrate to the hop, although some of the lice have been known to remain on the plum until cold weather. On the hop the winged migrants start colonies of wingless viviparous females. These forms reproduce asexually until the approach of cold weather, when winged females are produced which migrate back to the plum and there give birth to a small number of young that become wingless egg-laying females. The latter are fertilized by male migrants from the hops and then deposit the winter eggs on the terminal twigs. It has been shown in California that this plant-louse frequently, if not normally, remains on the hop plant throughout the year. The wingless forms are a uniform yellowish-green, while those having wings have black markings on the thorax and abdomen.

Treatment.

Unless occurring in excessive abundance this species will rarely require treatment on plum. Kerosene emulsion, whale-oil soap or tobacco extracts give satisfactory results, if applied before the curling of the leaves makes it impossible to hit the insects.

REFERENCES

Riley, Rept. U. S. Comm. Agr. for 1888, pp. 93–111.
Cal. Agr. Exp. Sta. Bull. 160. 1904.
Col. Agr. Exp. Sta. Bull. 133, pp. 40–41. 1908.

THE PLUM PLANT-LOUSE

Myzus mahaleb Fonscolombe

This yellowish-green plant-louse closely resembles the preceding species with which it has often been confused. It belongs to a different genus and may be distinguished by the structure of the head and antennæ; the frontal tubercles are not prolonged into a slender tooth and the first antennal joint is without the blunt tooth present in that species. It is not unusual to find both forms infesting the same tree.

The life history of the two species is very similar, but this form does not migrate to the hop and is able to subsist during the summer on a large number of plants among which may be mentioned the pear, sunflower, dock, kohlrabi, chrysanthemum, shepherd's purse, portulaca, etc. The winter eggs and spring generations occur on both plum and peach.

This plant-louse can be controlled by timely spraying with "Black Leaf 40" tobacco extract, $\frac{3}{4}$ pint in 100 gallons of water, adding 3 or 4 pounds of soap to make the liquid stick and spread better. Effective work can also be done with whale-oil soap solution or with kerosene emulsion.

REFERENCE

Pergande, U. S. Bur. Ent. Bull. 7, pp. 52–59. 1897.

s

The Mealy Plum Louse

Hyalopterus arundinis Fabricius

This light green plant-louse of the plum is generally distributed throughout the country wherever its food-plant is grown. It may be recognized by the three longitudinal darker green stripes above and by the fine white powdery covering of the body.

The black shining eggs are deposited in the fall around the buds on the smaller branches and hatch the following spring soon after the opening of the leaves. The first generation is composed entirely of wingless females known as stem-mothers. They give birth to a large number of living young, all females, which continue to reproduce in the same way for several generations. At first they are all wingless, but as the leaves become crowded and the food supply scanty part of each brood acquires wings and migrates to other plants until by the last of July practically all will have left the trees, but in some instances they have been known to remain on the plum throughout the season. The alternate food-plants are, as far as known, various species of grasses, particularly the Reed-grass (*Phragmites phragmites* Linn). In September the return migrants appear on the trees and give birth to winged males and wingless females. After mating the latter deposit the winter eggs around the buds.

Sometimes this plant-louse occurs in enormous numbers covering the entire under surface of the leaves, which turn yellowish and drop without becoming noticeably curled. The loss of foliage has been known to cause the falling of the fruit, but such severe attacks are unusual.

Treatment.

In case the plant-lice appear in sufficient number as to threaten injury to the crop it will pay to make a thorough spraying, using kerosene emulsion, or one of the tobacco extracts to which a small amount of soap has been added. Fortunately this species does not curl the leaves to any great extent, thus making it easier to do effective work.

REFERENCES

N. Y. (Geneva) Agr. Exp. Sta. Bull. 139. 1897.
Col. Agr. Exp. Sta. Bull. 133, p. 39. 1908.

THE RUSTY BROWN PLUM APHIS

Aphis setariæ Thomas

This dark rusty brown plant-louse occurs throughout the Southern states and is found as far north as Minnesota. It

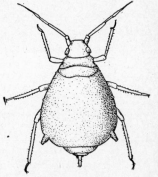

FIG. 226. — Stem-mother of the rusty brown plum aphis. Redrawn after Sanborn. Enlarged.

sometimes causes considerable loss to the plum crop, particularly in the South, and has been known to attack the peach, being especially troublesome on newly budded peach stock in nurseries. It is sometimes called the southern plum aphis under the scientific name of *Aphis scotti*. The general color of the body is rusty brown with the base of antennæ, tibiæ and tail white.

The dark brown shining winter eggs are attached to the smaller branches and hatch about the time the buds open. The first brood consists entirely of wingless females known as stem-mothers (Fig. 226) ; they give birth to living young which develop into wingless females. There are ten or more generations during the season. The earlier broods are wingless, but as the leaves become crowded winged individuals are produced which migrate to other plants, mostly grasses, and start new colonies. A few, however, may remain on the plum throughout the season. At the approach of cold weather winged forms are produced that return to the plum and there give birth to a brood of winged males and wingless females, both sexes being nearly black in color. The female deposits the winter eggs upon the bark of the smaller branches, usually near the buds.

Treatment.

The usual sprays for plant-lice are effective against this species, but as the infested leaves curl badly, the application should be made as soon as the insects appear. Thorough applications of lime-sulfur at the strength used against the San José scale are said to kill the eggs in Oklahoma.

REFERENCES

Scott, U. S. Bur. Ent. Bull. 31, pp. 56–59. 1902.
Ga. State Bd. Ent. Bull. 17, pp. 99–101. 1905.
Col. Agr. Exp. Sta. Bull. 133, p. 41. 1908.
Okl. Agr. Exp. Sta. Bull. 88. 1910.

The European Fruit-tree Scale

Aspidiotus ostreæformis Curtis

This scale insect closely resembles Putnam's scale and the cherry scale in size and general appearance and can be distinguished from them only by microscopic characters. The full-grown female scale is nearly circular, about $\frac{1}{12}$ of an inch in diameter, dark, ashy-gray in color and is usually easily distinguished from the San José scale by the exuvial spot, which, instead of being blackish and central, is dark orange colored and located a little to one side of the center. Probably introduced into New York State from Europe on orchard cuttings about 1870, this scale is now widely distributed in Canada and the northern United States.

The bark of plum, apricot, currant and soft maple trees are

sometimes incrusted and a few trees have been killed by the scale in this country. It also attacks the apple, pear, cherry and prune, besides several shade trees. It spreads slowly in orchards, and as there is apparently but one generation a year, at least in northern regions, it will never be as dangerous or destructive in orchards as the San José scale.

Only partly grown scales live through the winter, and these may reach maturity by the middle of June in New York. Some state that the mother scale insect gives birth to living young or is ovoviviparous, but we have seen eggs under the scales, and Reh records finding both eggs and living young under the scales in Europe, so the species may be partially oviparous also. Young are born over a period of several weeks and some of them often get on to the fruits, especially apples.

The lady-bird beetles, *Microweisea misella* and *Chilocorus bivulnerus*, feed upon this scale and parasites also destroy large numbers of them.

The lime-sulfur wash and other winter applications recommended for the San José scale will doubtless effectually control this European fruit-tree scale.

REFERENCE

U. S. Bur. Ent. Bull. 20, pp. 76–82. 1899.

THE EUROPEAN FRUIT LECANIUM

Lecanium corni Bouché

This large, brown, soft-bodied scale occurs on a large number of deciduous trees in both this country and Europe. Its most important hosts are the plum, peach, apricot, pear, apple, quince, currant, blackberry, mulberry, osage-orange, pecan, and among forest trees, elm, ash, basswood, etc. It has received a long list of scientific names, owing to slight variations in form and color assumed when growing on different plants. Although widely distributed throughout the United States and

Canada and infesting such a wide range of cultivated plants, it has become of economic importance only a few times and in restricted localities.

The most serious outbreak occurred in 1894 and 1895 in the plum orchards of western New York. The pest appeared suddenly in extraordinary numbers and excited widespread alarm among the fruit-growers

Fig. 227. — The European fruit Lecanium, eggs rolling out from beneath the female scale in June. Enlarged.

of that region. Fortunately, severe winter conditions, the attacks of parasites or some other cause so reduced their numbers that within a year or two the pest again subsided into comparative obscurity.

This scale insect has been studied principally as a plum pest. The winter is passed by the young, flat, spindle-shaped brown scales (Fig. 229) on the bark of the tree, mostly on the smaller branches. They are then about $\frac{1}{25}$ inch in length. Early in the spring, about April 1 in New York, they establish themselves mostly on the underside of the smaller branches (Fig. 228). The females grow rapidly and by the middle of May many are mature. They are then about $\frac{1}{8}$ or $\frac{3}{16}$ inch in length and remind one of small halved peas colored brown. The male

Fig. 228. — A plum branch infested with the European fruit Lecanium.

scale is much smaller, flatter and more elongate and is of a whitish color. About the time the females become mature the small, delicate, white-winged males appear and soon after mating perish. During the latter part of May or in June the female lays a large number of minute, white eggs which completely fill the cavity beneath the scale previously occupied by her body (Fig. 227). Over 2000 eggs have been counted beneath a single scale. They hatch in about a month and the minute young lice crawl out on to the leaves and establish themselves along the principal veins. Infested leaves become curled and turn yellowish, the tree makes but little growth and the fruit remains undersized or falls prematurely. The young scales collectively produce a great quantity of a clear sweet liquid known as honey-dew, which in the absence of heavy rains collects on the leaves and fruit and serves as a medium for the growth of a fungus which gives the tree a smutty, disgusting appearance.

Fig. 229. — Hibernating scales on plum of the European fruit Lecanium.

During September most of the scales abandon the leaves and seek winter quarters on the bark of the tree, mostly on the smaller branches, but a small percentage fall with the leaves and probably perish.

Treatment.

Commercial plum growers have found little difficulty in controlling this scale by one or two thorough applications of kerosene emulsion, diluted with 4 or 5 parts of water, applied while the trees are dormant. Miscible oils at the usual winter strength can also be used to advantage.

REFERENCES

Cornell Agr. Exp. Sta. Bull. 83. 1894.
Cornell Agr. Exp. Sta. Bull. 108, pp. 82–86. 1896.

N. Y. (Geneva), Agr. Exp. Sta. Rept. of 1895, pp. 574–595.   1896.
Sanders, Jour. Ec. Ent. II, pp. 443–445.   1909.

The plum Pulvinaria (*Pulvinaria amygdali* Cockerell) in general appearance closely resembles the cottony maple scale (p. 427).   It has been reported as attacking peach, apple, plum and prune.   It occurs in New Mexico, Georgia and California.

OTHER PLUM INSECTS

LESSER APPLE WORM : *apple*, p. 23.
GREEN FRUIT-WORMS : *apple*, p. 39.
BUD-MOTH : *apple*, p. 42.
CIGAR-CASE-BEARER : *apple*, p. 47.
FRUIT-TREE LEAF-ROLLER : *apple*, p. 62.
OBLIQUE-BANDED LEAF-ROLLER : *apple*, p. 65.
APPLE LEAF-SKELETONIZER : *apple*, p. 67.
LEAF-CRUMPLER : *apple*, p. 68.
CANKER-WORM : *apple*, p. 77.
WHITE-MARKED TUSSOCK-MOTH : *apple*, p. 100.
APPLE-TREE TENT-CATERPILLAR : *apple*, p. 112.
FOREST TENT-CATERPILLAR : *apple*, p. 119.
RED-HUMPED APPLE CATERPILLAR : *apple*, p. 125.
APPLE BUD-APHIS : *apple*, p. 151.
BUFFALO TREE-HOPPER : *apple*, p. 160.
SAN JOSÉ SCALE : *apple*, p. 162.
OYSTER-SHELL SCALE : *apple*, p. 171.
PUTNAM'S SCALE : *apple*, p. 179.
FLAT-HEADED APPLE-TREE BORER : *apple*, p. 194.
TWIG-PRUNER : *apple*, p. 200.
TWIG-GIRDLER : *apple*, p. 202.
FLEA-BEETLES : *apple*, p. 203.
CLOVER-MITE : *apple*, p. 206.
RING-LEGGED TREE-BUG : *apple*, p. 208.
EYE-SPOTTED APPLE-TWIG BORER : *apple*, p. 209.
NEW YORK WEEVIL : *apple*, p. 210.
SNOWY TREE-CRICKET : *apple*, p. 211.
PEAR THRIPS : *pear*, p. 223.
HOWARD SCALE : *pear*, p. 234.
EUROPEAN PEAR SCALE : *pear*, p. 234.
CHERRY SCALE : *cherry*, p. 312.

CHAPTER VIII

PEACH INSECTS

WHILE the peach easily succumbs to attacks of the San José scale, the loss occasioned by this insect is not as great as that produced by either the plum curculio or the peach-tree borers. The curculio is exceedingly destructive in the South, owing to the fact that it is an important factor in the distribution of brown rot spores. The control of the curculio on the peach is discussed on page 248. The foliage of the peach is very easily injured by spray mixtures and one must be especially careful in making applications of arsenicals.

THE PEACH-TREE BORER

Sanninoidea exitiosa Say

Wherever the peach is grown in the United States and Canada east of the Rocky Mountains it is subject to the attack of this native American borer, the larva of a beautiful steel-blue clear-wing moth. With the exception of the San José scale it is the most serious pest with which the northern peach grower has to contend, while in the South it is surpassed in importance only by the plum curculio. Quaintance in 1909 estimates the loss occasioned by peach-tree borers at 6 million dollars. Trees of all ages from nursery stock to the last relics of abandoned orchards are subject to attack; young trees are often girdled and killed outright, while older trees are so weakened that they are unable to produce good crops of fruit. Trees whose vitality has been reduced by borer attacks are particularly liable to infestation by bark-beetles, or shot-hole borers, which soon complete their destruction. The peach-tree borer also infests the wild and cultivated cherry, plum, prune, nectarine, apricot and certain ornamental shrubs.

The insect always passes the winter in the larval state. Throughout its range the great majority of the wintering larvæ are less than one half grown, although a very few may be nearly mature. Most of the larger larvæ hibernate in their burrows beneath the bark, but, as a rule, the smaller ones, those less than one half grown, pass the winter on the bark curled up under a thin silken covering or hibernaculum which protects them

FIG. 230. — The peach-tree borer, full-grown ($\times 3\frac{1}{4}$).

from the surrounding mass of gum. There are, however, exceptions in both cases; nearly full-grown larvæ are sometimes found in hibernaculums and some of the smaller ones hibernate in their burrows. In the extreme South the larvæ remain in their burrows, feeding nearly all winter, but in New York activity is not resumed until the last of April or in May. At that time the borers either continue the old burrows or start new ones and soon excavate a cavity from one half to an inch or more wide and two or three inches long just under the outer bark in the inner bark and sap wood. A large quantity of gum exudes from the burrows and collects in conspicuous masses on the ground at the base of the tree. We have seen a peach tree, only one and one half inches in diameter, support nine borers nearly to maturity in a single season, and it is not un-

Fig. 231. — Peach-tree borers in their burrows; cocoon above with an empty pupa skin projecting ($\times 1\frac{1}{4}$).

common to find old trees in neglected orchards infested with 40 or 50 borers.

Usually the larvæ confine their work to the trunk or roots of the tree a short distance below the surface of the soil, but they are sometimes found six or eight inches under ground. The full-grown larva (Fig. 230) is about an inch in length and of a very light yellow color; the head is dark reddish-brown, and the thoracic and anal shields are light brown. The body is sparsely clothed with brownish hairs arising from smooth, slightly elevated tubercles.

FIG. 232. — Cocoons of the peach-tree borer with empty pupa skins projecting.

When full-grown the larva usually leaves its burrow and incloses itself in a rough, brown, elongate oval cocoon composed of silk in which are incorporated particles of bark and excrement (Figs. 231 and 232). The cocoons are usually attached to the bark of the tree at or near the surface of the ground, but many are found lying loosely in the soil; this is more commonly the case in the South. In three to five days after building the cocoon the larva transforms into a dark brown pupa about $\frac{3}{4}$ inch in length (Fig. 233). In three or four weeks the pupa works itself halfway out of the cocoon and the steel-blue, clear-winged moth escapes.

FIG. 233. — Male pupa of the peach-tree borer ($\times 3\frac{1}{3}$).

The male and female moths differ strikingly in color (Fig.

234) ; in the male the four wings are transparent, with the veins, margins and fringes steel-blue, and the abdominal segments are steel-blue narrowly fringed with yellow behind ; in the female the front wings are opaque, covered with steel-blue scales, the hind wings are transparent over only about half their area and the abdomen has a broad, orange-colored band extending nearly around the body on the fourth or on both the fourth and fifth segments.

Fig. 234. — Male and female moths of the peach-tree borer.

North of latitude 40 to 42 degrees the females, as a rule, have the band covering both segments, while south of that line it is confined to the fourth segment. Intergrading specimens are sometimes found. The moths are most active during the heat of the day and when flying in the bright sunshine are easily mistaken for wasps by the casual observer.

Fig. 235. — Eggs of the peach-tree borer, enlarged.

Mating takes place very soon after the emergence of the moths and the female immediately begins to deposit her eggs singly or in small groups on the trunk of the tree, usually near the base. The egg is about $\frac{1}{50}$ inch in length, chestnut or reddish-brown in color, ellipsoidal in form, slightly flattened at the sides, truncate and slightly hollowed at the tip and beautifully sculptured, as shown in Figure 235. It has been shown by dissection of the moths that each female is capable of laying from 200 to 600 eggs. They hatch in 9 or 10 days and the young larvæ soon start their burrows in the soft bark.

The exact time of the emergence of the moths and the deposition of the eggs is a matter of great practical importance, and although it has received considerable attention by entomological workers, the data now available are not as complete as might be wished. In New York the moths have been known to emerge as early as June 12 and as late as the last of September, but most of them appear from July 1 to August 15. In New Jersey the period of maximum emergence is from June 15 to September 15; at Washington, D. C., it is practically the same, although a few moths may emerge in late May. In South Carolina most

FIG. 236. — Empty pupa skin of the lesser peach-tree borer projecting from its burrow in a plum branch.

of the moths emerge between July 20 and September 20, after which date only a few stragglers are seen. In Georgia and generally throughout the South a few moths may emerge as early as April or May, but the majority appear between August 1 and October 1. We would naturally expect the larvæ to mature and give rise to the adults earlier in the warm climate of the Southern states than in New York or New Jersey with their shorter growing season, but it is now well proved that such is not the case. The cause of this peculiar phenomenon is unknown.

Treatment.

The best results in the control of the peach-tree borer in commercial orchards are, as a rule, obtained by digging out the borers with a knife or some similar instrument, after which the trunk is treated with some good protective wash and the earth mounded up around the tree to a height of six to eight inches. Some successful growers rely entirely on the digging-out and mounding methods and omit the wash. Where the pest is at all troublesome the trees should be gone over carefully twice a year; once as late as convenient in the fall and again the first part of June. In digging out the borers the earth is first removed from around the base of the tree to a depth of four or five inches when the larger burrows will be indicated by conspicuous masses of gum. By scraping the bark with a knife or brush most of the smaller ones can be easily located. Particularly in the fall many of the borers are on the surface of the bark, covered merely by a mass of gum, where they are easily found and destroyed. To get at the larger borers in their burrows in the bark and sapwood considerable cutting may be necessary, but if it is done carefully and mostly in the direction of the grain of the wood, the wound soon heals and little or no injury is done the tree.

After the borers have been dug out in June the earth should be replaced at once and mounded up around the trunk to a height of eight or ten inches. This forces the moths to deposit their eggs higher on the trunk and causes the larvæ to enter the bark farther from the roots, where it is easier to locate and destroy them. The combination of the digging-out and mounding methods is the cheapest and most practicable way of controlling the peach-tree borer. The number of borers can be kept below the danger limit by this system alone if the work be thoroughly and regularly done, and if there are no neglected orchards near by to furnish moths for a constant reinfestation.

After the removal of the borers in the fall there is nothing to be gained by applying washes or wrappers, but many growers believe it pays to make such applications in June before mounding the earth around the trunks. At least fifty different kinds of washes have been suggested for preventing the attacks of the peach-tree borer in the century or more that American fruit-growers have been fighting this pest; some have been found injurious to the tree and many others practically worthless; most of these have been eliminated so that now only a very few are in common use. In general, washes are less effective than one would expect because it is very difficult to cover the rough bark of the tree trunk thoroughly enough to fill all the minute cracks and crevices through which the young larvæ usually gain entrance. Furthermore, the growth of the tree causes the wash to crack, thus exposing a fresh surface to the young borers. Many washes contain a poison, Paris green or arsenate of lead, intended to poison the borers as they gnaw through the protective coat, but they are of doubtful value. The young larvæ are very active and will usually be able to find some unprotected crevice before beginning to feed.

In an extensive series of experiments conducted in New York in 1895–1900 and also in the hands of some commercial growers gas tar as a protective wash has given excellent results, but as others have found it injurious to the trees under certain conditions it should be used with caution. It should not be applied to trees the first year after planting, before they are thoroughly established, nor should it be used in the fall. If applied to healthy trees in the spring while the vegetative growth is active, there is little danger of injury. A wash much used in commercial orchards in Georgia consists of 2 quarts soap, $\frac{1}{2}$ pint crude carbolic acid, 2 ounces Paris green, all mixed in a pail of water, to which enough lime and clay have been added to make a thin paste. In New Jersey a wash is strongly recommended composed of 1 pound arsenate of lead in 5 gallons lime-

T

sulfur at ordinary winter strength to which has been added an excess of lime.

Various kinds of wrappers for the prevention of the entrance of the borers were formerly much used, but are now generally discarded because of their tendency to render the bark tender. Where conditions are such as to warrant their use wrappers made of several layers of newspapers or cheap brown paper tied tightly around the trunk and extending from the roots to a height of about two feet give fairly good protection at a minimum of expense and labor. Tarred paper is frequently recommended, but it is no more effective than the cheaper newspaper wrappers. Tobacco stems wound tightly around the trunk from the roots to a little above the surface of the ground kept out from two thirds to five sixths of the borers in the New York experiment mentioned above. They are cheaply obtained at cigar factories and are worthy of further trial. Wrappers to do the most good should be in place and kept intact throughout the egg-laying period, which varies in different parts of the country, as stated above. To be most effective washes and wrappers should always be used in combination with the digging-out and mounding methods.

Wire cages loosely encircling the trunk and tightly closed at the top with cotton were strongly recommended in Missouri some years ago, but further experiments have shown that they are practically worthless, although theoretically they should give perfect protection. Actually more borers were found in trees fitted with these devices than in those entirely unprotected.

REFERENCES

N. J. Agr. Exp. Sta. Bull. 128. 1898.
Cornell Agr. Exp. Sta. Bull. 176. 1899.
Cornell Agr. Exp. Sta. Bull. 192. 1901.
Georgia Agr. Exp. Sta. Bull. 73. 1906.
N. J. Agr. Exp. Sta. Bull. 235. 1911.

The Pacific Peach-tree Borer

Sanninoidea opalescens Henry Edwards

This insect, closely related to the peach-tree borer of the Eastern states, attacks the peach, almond, apricot, prune and cherry on the Pacific Coast. Its habits and mode of attack are very similar to those of the Eastern form. Nearly all sizes of larvæ are found in the trees in the winter; these mature and transform to moths at different times throughout the growing season, thus giving a long period during which the eggs are deposited. The male moth has the transverse mark and outer margin of the front wings broader than in the Eastern species; in the female the fore wings are opaque, the hind wings transparent and the abdomen is entirely blue or green-black, without any yellow band.

This pest may be controlled by digging out the larva during the winter or early spring. To prevent the entrance of the newly hatched larvæ a protective wash should be applied to the trunk, from the roots to a height of 18 inches from the surface of the ground. Excellent results have been obtained by using a wash made according to the following formula, and known as the lime-crude-oil mixture: place about 50 pounds of rock lime in a barrel and slake with 10 or 15 gallons of warm water; while the lime is boiling, slowly pour in 6 or 8 gallons of heavy crude oil, and stir thoroughly. Add enough water to make the whole a heavy paste. Asphaltum has also been found a safe and efficient protection.

The bisulphide of carbon treatment, strongly recommended in California a few years ago, is now generally discarded as being too dangerous and expensive.

References

Cal. Agr. Exp. Sta. Bull. 143. 1902.
U. S. Bur. Ent. Bull. 97, Pt. IV. 1911.

THE LESSER PEACH-TREE BORER

Sesia pictipes Grote and Robinson

This insect is also closely related to the peach-tree borer, with which it has sometimes been confused. Besides the peach it also attacks plum, cherry, June berry, beach plum and chestnut. It is widely distributed throughout the United States and Canada.

Unlike the peach-tree borer the larvæ do not confine their attacks to the crown but more often occur on the trunk and larger branches. They usually infest old trees with rough bark, and most of the burrows are found in the vicinity of wounds. The pinkish or translucent larvæ very closely resemble those of the peach-tree borer, but are somewhat smaller. When full-grown the larva eats out an exit hole nearly through the bark, and in a suitable cavity beneath constructs an oval cocoon of bits of bark and frass bound together with silken thread. A few days after the construction of the cocoon the larva changes into a brownish-yellow, spindle-shaped pupa about $\frac{3}{5}$ inch in length. The moth emerges in about a month or a little less in the South, leaving the empty pupal skin projecting from the burrow (Fig. 236). Both sexes of the moths resemble the male of the peach-tree borer, having all four wings transparent. Larvæ of all stages except the first may be found during the winter. These mature at different times throughout the summer and give a continuous supply of moths. In the South there are two generations annually, while in the North there is probably only one.

Treatment.

This borer rarely attacks perfectly sound uninjured trees and is of little economic importance in commercial orchards which receive good care. It is best controlled by digging out the larvæ, after which the wounds should be treated with some protective wash. The work may be done in connection with the treatment for the peach-tree borer.

REFERENCE

U. S. Bur. Ent. Bull. 68, Pt. IV.   1907.

THE FRUIT-TREE BARK-BEETLE

Scolytus rugulosus Ratzeburg

This European bark-beetle was first noticed in this country in 1877 in the vicinity of Elmira, N.Y.   It now occurs in Canada,

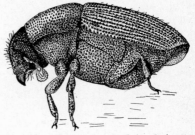

and is generally distributed over the Eastern states as far South as Alabama and Georgia.   It breeds freely in the peach, plum, cherry, apricot and apple and will also attack quince, mountain ash, June berry and chokecherry.   Its presence is indicated by numerous

FIG. 237.— The fruit-tree bark-beetle (×20).

small exit holes in the bark, about $\frac{1}{16}$ inch in diameter, hence its common name of shot-hole borer.   In stone fruits its work is rendered conspicuous by the gum which oozes from the burrows and hangs in unsightly masses from the branches (Fig. 243).

FIG. 238. — Brood chamber and larval burrows of the fruit-tree bark-beetle.

The adult, or beetle, is about $\frac{1}{10}$ inch in length and of a dark brown color, except parts of the legs and the tips of the wing covers, which are dull reddish

FIG. 239. — Peach branch with bark removed to show the burrows of the fruit-tree bark-beetle.

(Fig. 237). They emerge in early spring, and the female immediately seeks out a suitable branch in which to deposit her eggs. Having selected the place, she burrows directly to the sapwood, and there partly in the wood and partly in the bark constructs an egg chamber which usually runs lengthwise of the branch and is from $\frac{1}{2}$ to 2 inches long (Fig. 238). From time to time as the burrow advances she lays her minute, delicate white eggs close together in a single row on each side of the chamber, gluing them in place with a gummy secretion. It takes her not far from a week to complete the egg chamber, and as the eggs hatch in about three days those first laid hatch before the last are deposited. On hatching, the minute, whitish grubs start their burrows at right angles to the egg chamber, but soon change the direction, so that by the time they are full-grown most of the burrows are running lengthwise of the branch (Fig. 239). The full-grown larvæ are about $\frac{3}{16}$ inch in length, whitish in color, with brown mouth parts; the anterior segments are considerably enlarged, and the hind end of the body is obtusely rounded. They become grown in about 20 days, and then excavate a narrow cavity in the sapwood about $\frac{1}{8}$ inch deep, plug the opening with sawdust and transform to pupæ within (Fig. 241). In about 10 days the beetles gnaw their way to the surface. The complete life cycle requires from four to six weeks. The separate broods greatly overlap, making it very difficult to determine accurately the number of generations a year. The evidence

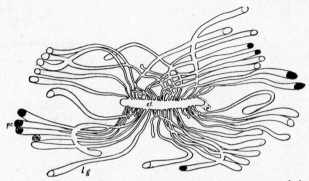

FIG. 240. — Burrows of the fruit-tree bark-beetle; (e) entrance hole, (e.t.) egg tunnel cut by adult, (lg) larval gallery, (p.c.) plugged entrance to pupal cell. Redrawn after Swaine.

available goes to show that in the North there are probably two generations, while in the South there are three, four or even more. The winter is passed as larvæ either partially grown in the burrows or full-grown in the pupal chamber.

These bark beetles do not seem to be able to breed either in healthy, strongly growing trees or in entirely dead dry branches. They prefer trees that have been weakened from injury, lack of care, or from some other cause. Still when very abundant they will attempt to enter perfectly healthy trees, selecting those parts where the vitality is lowest.

FIG. 241. — Section through bark and wood of apple branch infested with the fruit-tree bark-beetle showing its burrow filled with frass and the larva in its pupal cell. Enlarged.

In the case of stone fruits they are either driven out or killed by the copious flow of

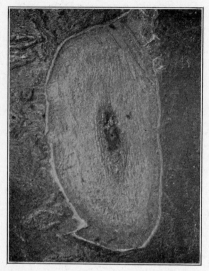

FIG. 242. — Fruit-tree bark-beetle killed by gum while attempting to enter a healthy peach tree for egg-laying. Enlarged.

sap into their burrows (Fig. 242), but continued attacks will in time so weaken the tree that the females are able to gain a foothold and deposit their eggs. On cherries a peculiar injury is often noticed in midsummer. The short spurs bearing the clusters of leaves are attacked by a beetle, which eats out a short burrow apparently for food only, since eggs are never found in these burrows, and they are soon deserted. The dead leaves remain on the branch for some time and call attention to the injury.

Treatment.

To avoid an infestation by bark beetles the trees should be kept in a vigorous condition by proper cultivation, pruning and

FIG. 243. — Gum exuding from a peach branch infested with the fruit-tree bark-beetle.

spraying, for the beetles are not able to breed in healthy, strongly growing trees or branches. All trees in too poor condition to be of commercial value should be removed before they become centers of infestation. Near-by abandoned orchards, brush piles, old neglected cherry trees, and chokecherry trees along roadsides and fences frequently harbor the pests in countless numbers and should be kept under close observation or destroyed. We have seen a thrifty young apple orchard severely attacked by beetles that come from a large pile of fire wood from an old apple orchard that had been cut down. When an orchard has become badly infested severely injured trees and branches should be removed and burned before the beetles have had a chance to escape. Slightly infested trees may sometimes be saved by a rather severe pruning, thorough cultivation and the application of some quick-acting fertilizer. After the beetles have entered the bark there is no practicable way of reaching them with an insecticide. A 12 per cent emulsion of avenarius carbolineum is said to kill the beetles in their burrows. Certain deterrent washes, however, have been shown to possess considerable protective value when applied in the spring just before the beetles appear and repeated once or twice during the season. A stiff whitewash applied to the trunk and larger branches has given good results; its efficiency is increased by the addition of one gallon of chloronaphtholeum or avenarius carbolineum to each 50 gallons of the wash, and the addition of one fourth pound of salt to each pail of the wash will add to its sticking qualities. In commercial peach orchards in Georgia fairly good protection has been procured at a reasonable cost by the use of lime-sulfur at the strength used against scale, the application being made just before the appearance of the beetles in the spring.

REFERENCES

Ill. Agr. Exp. Sta. Bull. 15. 1891.
N. J. Agr. Exp. Sta. 15th Ann. Rept. for 1894, pp. 565–572.
U. S. Bur. Ent. Circ. 29 (Revised edition), 1903.
Ohio Agr. Exp. Sta. Circ. 140. 1913.

The Peach Bark-beetle

Phlœotribus liminaris Harris

In habits and in the nature of its injuries this species is very similar to the foregoing, but its range of food plants is more restricted, its attacks being confined principally to the peach and cherry. It is a native of this country and occurs in Canada and the Eastern states as far south as North Carolina.

Unlike the preceding species it hibernates in the beetle stage either in the pupal cavities or in specially constructed hibernation chambers excavated in the bark of healthy trees. The small brownish beetles (Fig. 244) emerge in early spring and at once begin to excavate their egg burrows under the bark of weakened or dying trees. These burrows usually

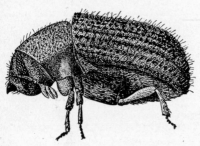

Fig. 244. — The peach bark-beetle (× 22).

run transversely around the branch and are distinguished from those of the preceding species by having a short side branch (Fig. 245) which with the short tunnel running to the entrance hole in the bark makes a Y-shaped end to the main egg burrow. The side branch enables the female to turn around in the burrow and is also occupied by the male at the time of mating. The minute white eggs are deposited in small niches in the walls of the main burrow and are covered with sawdust. Each female may lay from 80 to 160 eggs. Those of the first brood hatch in about 20 days, while those of the summer generation require only about half as long. The young grubs burrow away from the egg chamber at right angles and generally follow the grain of the wood, but gradually diverge as they advance. In 25 or 30 days the full-grown grubs transform to pupæ in the enlarged ends of the burrows near the outer surface of the bark. From 4 to 6 days are spent in the pupal stage, but as a rule the beetles do not emerge until a week or two later. In northern Ohio there are two generations a year, the beetles appearing in greatest numbers in March and early April, in July and again in October, when they go into hibernation. The broods overlap, so that after July all stages are present under the bark until cold weather.

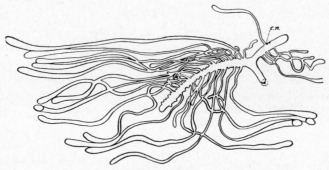

Fig. 245. — Burrows of the peach bark-beetle; (e. n.) egg nitch. Redrawn after Swaine.

The beetles may be distinguished from the preceding species by the following points : the club of the antenna is lamellate ; when viewed from the side the venter of the abdomen appears nearly straight, not turned abruptly upward and the pronotum is bent strongly downward, so that the head is scarcely visible from above (Fig. 244).

Treatment.

The measures suggested against the fruit-tree bark-beetle are also applicable to the present species. From experiments conducted in Ohio it is recommended that, in order to keep out the greatest number of beetles, the whitewash should be ap-

plied three times : in late March or early April, about the middle
of July and in late September or October.

REFERENCE

U. S. Bur. Ent. Bull. 68, Pt. IX.   1909.

THE PEACH TWIG-BORER

Anarsia lineatella Zeller

This European peach pest is now generally distributed
throughout the United States and Canada wherever its host

plant is cultivated.
It is sometimes
troublesome in the
Eastern states and
has become a serious
pest on the Pacific
Coast, where it is

FIG. 246. — Moth of the peach twig-borer (× 5).

estimated that its attacks cause a loss of about one fourth
of the peach crop in certain regions.   The over-wintering
larvæ burrow into the tender shoots in early spring, cause
them to die and give the tree the appearance of having
been scorched by fire ; the summer generations likewise burrow
in the new growth but also attack the fruit, particularly of
late varieties.

The insect hibernates as a small larva, about $\frac{1}{16}$ inch in
length, in a silk-lined cavity just beneath the outer bark, usually
in the crotch at the base of the new growth.   The location of
the hibernating cavities is indicated by the small, reddish-brown
mounds of bits of bark webbed together with silk thrown out
of the cavity.   In early spring the larva enlarges the cavity
by feeding on the surrounding tissues and reaches the surface
in 10 to 14 days.   The larvæ then attack the young growth, bur-

rowing into the pith of the tender shoots, which soon wilt and die. The same larva usually attacks a number of twigs in succession thus causing an amount of injury out of proportion to the quantity of tissue consumed. Three or four larvæ have been known to kill a three-year-old tree by destroying all the new growth.

When full-grown, the reddish-brown larvæ, about $\frac{2}{5}$ to $\frac{1}{2}$ inch in length, crawl to the larger branches or trunk, where within the curled flakes of the outer bark they construct very loose cocoons consisting of only a few threads of silk. The pupa varies from light to dark yellow in color and is about $\frac{1}{4}$ inch in length. The pupal period lasts from 10 to 12 days. The inconspicuous, steel-gray moths (Fig. 246), with an expanse of about $\frac{1}{2}$ inch, deposit their white or yellowish elongate oval eggs on the bark of the new twigs near the base of the leaves. These eggs hatch in about ten days, and the second brood of caterpillars attack the tips of the growing branches during the latter part of May in California. After about 20 days, or early in June, they leave the burrows in the tips of the young shoots and attack the fruit if present. Most of these larvæ enter the fruit at the stem end along the suture and eat out a considerable cavity in the flesh, which becomes filled with excrement and gum; sometimes they burrow next the stone, or in case the pit is split they may devour the kernel. The second-brood larvæ become mature during July and August and pupate in very slight cocoons in the hollow at the stem end of the peach. The third brood of moths deposit their eggs on the fruit on the edge of the depression around the insertion of the stem. Soon after hatching, the third-brood larvæ enter the fruit and feed within until full-grown, when they eat their way out and pupate at the base of the fruit, as in the case of the second brood. Moths of the fourth brood begin to emerge by the middle of August, so that they become mixed with belated individuals of the third brood. These moths of the last brood deposit their eggs in late August and September in cracks and rough places in the bark. These eggs hatch in about five days, and the young larvæ soon excavate small cavities in the bark at the base of the new growth in which to pass the winter.

Treatment.

Extensive experiments in California have shown that the peach twig-borer may be satisfactorily controlled by thorough and timely spraying with the lime-sulfur wash. The application should be made just after the buds begin to swell and may be continued until the first blossoms appear. This is the period at which the young larvæ are leaving their winter quarters and are most easily reached by the spray. Kerosene or distillate emulsion may be used at this time, but is somewhat less effective and is more likely to cause injury. Winter applications of either the emulsions or the lime-sulfur wash are of little value in the control of this insect, because at that time the larvæ are out of harm's way in their hibernating burrows in the bark. It was formerly supposed that the oil would be absorbed by the frass, penetrate the burrow and kill the larva, but later work has shown that it penetrates very slowly if at all, and the practice is now generally discarded. In case the early application of the lime-sulfur wash has been omitted it will pay to spray the trunks and larger branches with kerosene or distillate emulsion in late spring to kill the first-brood pupæ in their flimsy cocoons in the curls of bark. This treatment cannot be relied upon to control the pest, but may be used to supplement the use of the lime-sulfur wash when for any reason it may seem to have been ineffective.

REFERENCES

U. S. Dept. Agr. Farm. Bull. 80. 1898.
Cal. Agr. Exp. Sta. Bull. 144. 1902.

Cenopis diluticostana Waslingham sometimes causes an injury to peaches very similar to the work of the peach twig-borer.

17th Rept. N. Y. State Ent., p. 736. 1901.

THE STRIPED PEACH WORM

Gelechia confusella Chambers

In restricted localities in Michigan peaches are sometimes attacked by small, dirty, yellowish-white caterpillars marked on the back and sides by six longitudinal, reddish-brown stripes; when full-grown they are about ⅜ inch in length. They appear in July and again in September and feed upon the leaves, which they web together into loose nests. When full-grown, they transform to pupæ within the nest. The winter is passed in the pupal state. The moth has an expanse of about ⅜ inch; the front wings are almost black with a purplish gloss; the hind wings are cinereous.

When only a few trees are infested, it will pay to cut out and burn the nests. In larger orchards the first-brood caterpillars can be destroyed when they first appear by spraying with arsenate of lead, 4 pounds to 100 gallons of water, to which 4 pounds of lime should be added to prevent burning of the foliage.

REFERENCES

Mich. Agr. Exp. Sta. Bull. 175, pp. 347–349. 1899.
Mich. Agr. Exp. Sta. Sp. Bull. 24, pp. 57. 1904.

THE PEACH SAWFLY

Pamphilius persicus MacGillivray

This insect has recently become locally troublesome in peach orchards in Connecticut; it also occurs in Nebraska. The adult sawflies are black with yellow markings on the head, thorax and antennæ, and with the abdomen behind the basal plates rufous; they are about ⅜ inch in length. They emerge from the ground in late May or early June, and the female deposits her pearly white, elongate eggs along the midrib on the underside of the leaf. The eggs hatch in about a week and the larva first eats out a narrow strip of the leaf from the edge towards the center and then rolls over a portion of the leaf, making a case within which it remains during the day, feeding mostly at night. The foliage of badly infested trees presents a characteristic shredded appearance.

The larvæ grow rapidly and become full-grown in about 10 days; they are then about $\frac{5}{8}$ inch in length and are of a pale bluish-green color. When mature they enter the ground to a depth of three to six inches, where they remain curled up in small round earthen cells about $\frac{1}{4}$ inch in diameter until the following spring. The transformation to the naked greenish pupa takes place in late May and early June and the adults appear about two weeks later. There is only one generation annually.

Treatment.

This pest has been satisfactorily controlled in commercial orchards by thorough spraying, just after the hatching of the eggs, with arsenate of lead, 5 to 6 pounds to 100 gallons of water. This is doubtless stronger than necessary; 4 pounds to 100 gallons would probably be just as efficient and safer for the foliage.

REFERENCE

Conn. Agr. Exp. Sta. Rept. for 1907, pp. 285–300. 1908.

Another species of sawfly (*Caliroa amygdalina* Rohwer) has been reported as attacking the peach and plum in Louisiana. The adult sawflies appear in March or April. The female inserts her semi-transparent flattened eggs into the tissue of the leaf from the upper surface, and they lie next to the lower epidermis; they hatch in four to six days. The tad-pole shaped larvæ are smooth and shining, and until the last stage are covered with a viscid slime. In feeding they skeletonize the leaf in the manner of the pear slug. They become full grown in 9 to 10 days, and then enter the ground to the depth of a few inches, where they transform to pupæ inside a tough brownish cocoon. The adult sawflies emerge in about nine days. The whole life cycle is completed in less than a month and there are several generations annually. Doubtless this insect can be controlled by the same measures as are recommended against the pear slug.

REFERENCES

La. Agr. Exp. Sta. Bull. 48, pp. 142–145. 1897.
U. S. Bur. Ent. Bull. 97, Pt. V. 1911.

THE BLACK PEACH APHIS

Aphis persicæ-niger Smith

This dark brown or black plant-louse is a serious enemy of the peach in certain parts of the Eastern states, particularly in Delaware, Maryland, New Jersey and Virginia. It occurs in Colorado and California and has been reported from Ontario, Canada. It is a native insect, and the wild plum was probably its original host plant.

This plant-louse feeds throughout the year upon the roots of the tree and during spring and early summer is also found upon the tender twigs and leaves. It breeds agamically all the year round, and no males or eggs have been found. Only the wingless forms occur upon the roots, where they breed freely, and are to be found at all seasons of the year, often a foot or more beneath the surface even in stiff clay soils, although sandy soils are preferred. Early in spring, as soon as the buds begin to swell, some of these root forms make their way to the surface and establish colonies on the new growth. They increase rapidly, so that in a few weeks the tender twigs and even the leaves become entirely covered by masses of the dark brown or black lice. As their feeding quarters become crowded winged forms (Fig. 248) develop and fly to other trees, where they es-

U

tablish new colonies. In midsummer most of the lice leave the branches and migrate to the roots, but usually a few remain above ground until winter. The full-grown wingless form

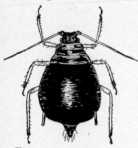

(Fig. 247) is a little less than $\frac{1}{10}$ inch in length, nearly black with portions of the legs yellowish. The winged form is a little longer and more slender; the young resemble the adults, but are lighter in color.

The greater part of the injury to the trees is caused by the underground form, although when very abundant the aërial forms may kill young trees. The injury caused by the root form is often mistaken for peach yellows or attributed to some

Fig. 247. — The black peach aphis, adult apterous viviparous female. Enlarged. Redrawn after Miss M. A. Palmer.

other cause, as winter injury, poor soil or the use of winter insecticides. The trees are stunted, do not put out the proper new growth and the foliage takes on a yellowish sickly appearance. Nursery stock is very liable to infestation, especially when grown on light sandy soil. The lice remain on the roots at the time of digging and are thus introduced into new orchards. Trees are most susceptible to injury the first and second years after planting. Even if not killed outright their vitality may be so weakened that they never

Fig. 248. — The black peach aphis, winged female. Enlarged. Redrawn after Miss M. A. Palmer.

fully recover, but always remain undersized and incapable of bearing full crops of fruit.

Treatment.

To avoid introducing the pest into new orchards set only

nursery stock which has been properly fumigated with hydro-cyanic acid gas. Where this is impossible the roots should be cleaned of dirt and immersed for a few minutes in a strong tobacco decoction to kill any lice which may have remained upon them. When trees are found to be infested after they are planted remove the earth about the tree to a depth of a few inches and apply a pound or so of tobacco dust and replace the earth. The tobacco acts both as an insecticide and as a ferti-lizer, killing some of the lice and helping the tree to outgrow the attack. The orchard should be kept under thorough cultiva-tion and supplied with proper plant-food to promote a strong, vigorous growth. If the aërial forms, on the new growth, be-come troublesome at any time, they can be controlled easily by thorough spraying with kerosene emulsion, whale-oil soap, 1 pound in 6 gallons of water, or tobacco extract combined with soap solution.

REFERENCES

N. J. Agr. Exp. Sta. Bull. 72, pp. 20–23. 1890.
Cornell Agr. Exp. Sta. Bull. 49, pp. 325–331. 1892.
U. S. Dept. Agr. Yearbook, 1905, pp. 342–344. 1906.
N. J. Agr. Exp. Sta. Bull. 235, pp. 32–35. 1911.

THE GREEN PEACH APHIS

Myzus persicæ Sulzer

The leaves, blossoms and young fruit of the peach are often attacked by a pale green or greenish-yellow plant-louse which sometimes causes serious injury to the crop. This European insect was introduced into this country many years ago and is now generally distributed throughout the United States and Canada. It has a wide range of food-plants, including, besides a number of deciduous fruits, the orange, many garden, truck and ornamental plants and numerous weeds; it is also a green-house pest. Until recently the form infesting garden and green-house plants has been known as *Rhopalosiphum dianthi* Shrank.

The black, shining winter eggs of the insect are found in the axils of the buds and in the crevices of the bark on the peach, plum, apricot, nectarine and cherry. The eggs hatch in early spring so that the pinkish stem-mothers are nearly full-grown by the time the blossoms open and soon begin to give birth to living young. These forms resemble the stem-mother in form, but are pale yellowish-green and usually have three longitudinal lines of darker green on the abdomen. A few of the second and nearly all of the third generation acquire wings and fly away to other plants. Among the many summer food-plants the most important are cabbage, cauliflower, radish, turnip and potato, while spinach, cucumber, tomato, egg plant, lettuce and celery are sometimes seriously injured. Violets, roses and carnations are particularly liable to infestation when grown under glass. During

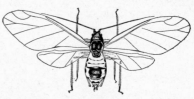

FIG. 249. — The green peach aphis, fall migrant. Enlarged. Redrawn after Miss M. A. Palmer.

the summer winged forms are produced from time to time as the feeding quarters become crowded. At the approach of cold weather, winged females (Fig. 249) return to the peach, establish themselves along the veins on the underside of the leaves and give birth to true or sexual females. The latter are usually of a pinkish color similar to the stem-mothers of the preceding spring. After pairing with the winged males, they deposit their eggs in the axils of the buds and in crevices of the bark. The insect does not hibernate exclusively in the egg stage, but the wingless forms are frequently able to survive the winter under suitable protection in the vicinity of their herbaceous food-plants.

The injury to the peach is all done early in the season; the early broods attack the blossoms and frequently blight them;

the leaves become curled, turn yellow or red and in severe cases drop; even the young fruit itself is sometimes attacked, wilts and falls prematurely.

Treatment.

The green peach aphis when infesting the peach is not a difficult insect to control by the use of contact sprays, provided the application is made before the curling of the leaves makes it impossible to hit the lice. The eggs hatch rather early in the spring, and the stem-mothers collect around the swelling buds. Excellent results have been obtained in commercial orchards by spraying at this time, just before the opening of the buds, with 5 per cent kerosene emulsion, miscible oils or tobacco extracts to which soap has been added at the rate of 4 pounds in 100 gallons.

REFERENCES

Taylor, Jour. Ec. Ent. I, pp. 83–91. 1908.

Col. Agr. Exp. Sta. Bull. 133, pp. 32–37. 1908.

Va. Truck Exp. Sta. Bull. 2, pp. 30–32. 1909. As a spinach pest.

THE TERRAPIN SCALE

Lecanium nigrofasciatum Pergande

This native enemy of the peach and plum is generally distributed throughout the United States east of the Mississippi and also occurs in Minnesota, Missouri, Arkansas and Texas. In addition to the peach and plum, it attacks, among others, the apple, olive, maple, basswood, birch and blueberry. Previous to 1898 this species was confused with the European peach scale (*Lecanium persicæ* Fabricius). As a peach pest it has caused considerable loss in the orchards of southern Pennsylvania, New Jersey and Maryland. The injury to the tree itself is not serious, but the principal loss comes from the honeydew secreted by the scales. This sweet, sticky substance accumulates on the tree and serves as a medium for the growth of a sooty black fungus that renders the fruit practically unsalable. The winter is passed by the fertilized females on the smaller branches. These scales are about $\frac{1}{12}$ inch in length, nearly circular in outline and very convex with an elongate reddish area on the back and radiating black lines along the margin.

FIG. 250. — The terrapin scale on a peach branch. Enlarged.

The edge of the scale is slightly ridged (Fig. 250). Individuals vary greatly in color from nearly pure black to orange-red. Growth is resumed in early spring, and by the last of May the females are about $\frac{1}{8}$ inch in length. Eggs are formed, and the minute straw-colored flattened young escape from beneath the old scale and establish themselves on the leaves. Hatching continues for a period of about 6 weeks during June and July. Winged males appear during August and fertilize the females; the latter soon migrate to the bark of the smaller branches, where the winter is spent. There is only one generation a year.

Treatment.

Recent experiments in Maryland have shown that this scale may be readily controlled by thorough spraying with miscible oils at the usual dilution for dormant trees. When used in the fall there is danger of killing the fruit-buds, but if the treatment is deferred until spring, when the tree has resumed its activities, this difficulty can be in great measure obviated. The use of lime-sulfur for this scale is of little or no value. Summer applications directed against the newly hatched active young have been found impracticable in commercial practice. Although the young are easily killed by applications of kerosene emulsion or nicotine extracts, so many sprayings are necessary to cover the long hatching period of at least six weeks that the expense becomes prohibitive in large orchards.

References

Murtfeldt, U. S. Bur. Ent. Bull. 32, O. S., pp. 41–44. 1894.
Pergande, U. S. Bur. Ent. Bull. 18, pp. 26–29. 1898.
U. S. Bur. Ent. Circ. 88. 1907.
Md. Agr. Exp. Sta. Bull. 123, pp. 153–160. 1907.
Md. Agr. Exp. Sta. Bull. 149. 1910.

The European peach scale (*Lecanium persicæ* Fabricius), according to Sanders, has been reported in this country only from California under the name *Lecanium magnoliarum* Cockerell.

The earlier accounts of *L. persicæ* in our economic literature usually refer to the terrapin scale.

The White Peach Scale

Aulacaspis pentagona Targioni Tozzetti

In the Southern states and in California the peach and related fruits are subject to the attack of a whitish scale frequently known as the West Indian peach scale. This insect is widely distributed throughout the warmer regions of the world, occurring in Ceylon, India, China, the West Indies, South Africa, Brazil, etc. Its range of food-plants is extensive, but in this country it has become troublesome only on the peach, plum, prune, cherry and apricot. As a peach pest it has been known to be fully as injurious as the San José scale.

The full-grown female scale is about $\frac{1}{10}$ inch in diameter, dirty white in color and nearly circular in outline; the yellowish or brownish exuvium is not generally central, although it is sometimes nearly so. The elongate, pure white male scales have a tendency to occur in close clusters usually at the base of the branch or trunk. The winter is passed by the mature females; in the latitude of Washington they begin egg-laying about May 1. The eggs hatch in a few days and after a short period of active life the young settle down and the protective scale is formed. By the middle of June the females become mature, and about that time the red-bodied, transparent-winged males emerge. The second brood of eggs is laid the last of June, and a third at the end of August; females of the last brood become mature in late October, in which condition hibernation takes place. There are thus three generations at Washington, while in Florida and Georgia there are said to be four broods.

Treatment.

Lime-sulfur as used against the San José scale has been found satisfactory for the control of this pest. The use of dilute lime-sulfur for the control of brown rot would do much to prevent the young from establishing themselves on the bark. The young may also be killed by summer applications of whale-oil soap, kerosene emulsion or tobacco extracts.

REFERENCES

Riley and Howard, Insect Life, VI, pp. 287–295. 1894.
Fla. Agr. Exp. Sta. Bull. 61, pp. 492–498. 1902.

The Green June-beetle

Allorhina nitida Linnæus

In the Southern states and along the Atlantic coast as far north as Long Island these large velvety green beetles (Fig. 251) are sometimes very troublesome in orchards and vineyards. The females are an inch or less in length and usually have

Fig. 251. — The green June beetle resting on earthen cocoon. Knight photo.

the sides of the thorax and
wing-covers brownish-yel-
low; the males are smaller
and have the yellowish
markings more diffuse.

The beetles frequently
appear in large numbers
about the time the fruit is
ripening and cause much
damage by breaking the

Fig. 252. — The green June beetle, empty cocoon and one containing a newly transformed beetle.

skin of the fruit, after which they feed on the juicy pulp
within. The large, thick-bodied dirty white grubs (Fig. 253),
nearly two inches in length, from which the beetles develop,
live in the ground and feed on decaying vegetable matter, not
on the roots of living plants, as was formerly supposed. They
are especially abundant in heavily manured gardens, about

Fig. 253. — Full-grown grubs of the green
June beetle.

Fig. 254. — Pupa of the green
June beetle in its cocoon. Knight
photo.

manure heaps and in fields that have been heavily mulched
with straw. When they come to the surface, as is often the case
after heavy rains, they have the curious habit of crawling on
their back, advancing with a wave-like motion of the ridges
of the body, which are armed dorsally with short, stiff bristles.

The female beetle deposits her white, nearly spherical eggs

an inch or two in the ground.   The young grubs burrow through
the soil, feeding on the humus, and become only partly grown
by winter.   They complete their growth the following spring,
pupate in earthen cells (Figs. 252 and 254) in the ground, and the
beetles begin to emerge in early summer, becoming most abundant during August in the latitude of Kentucky.   They attack
peaches, pears, plums, grapes and other fruits.   Corn in the
milk is often injured.

Treatment.

As it is impracticable to use arsenical sprays on ripening
fruit, it has been suggested that the beetles might be attracted
to piles of overripe and decaying fruit beneath the trees and
there poisoned with Paris green or arsenate of lead.   In practice this method is of little value because the majority of the
beetles prefer to feed on the fruit still hanging on the tree.
Thorough and repeated hand-picking of the beetles as fast as
they appear is the safest and surest way of protecting a crop.
Manure piles should not be permitted in the vicinity of orchards
where the pest is troublesome, and the use of mineral fertilizers
is suggested as a means of decreasing the number of grubs.

REFERENCES

Md. Agr. Exp. Sta. Bull. 23, pp. 77–81.   1893.
U. S. Bur. Ent. Bull. 10, pp. 20–26.   1898.
Ky. Agr. Exp. Sta. Bull. 116, pp. 67–73.   1904.

In the Southwest a closely related beetle (*Allorhina mutabilis*
Gory) causes similar injury to ripening fruits.   As far as known,
its habits and life history do not materially differ from those of
the last species, and the same remedial measures are suggested.

REFERENCE

N. Mex. Agr. Exp. Sta. Bull. 5, p. 10.   1892.

The brown fruit-chafer (*Euphoria inda* Linnæus) and its
near relative (*E. melancholica* Gory) are both known to at-

tack ripening fruits. Of the life history of the latter, little is known. The adult of the former is a yellowish-brown beetle, ½ inch or more in length, with its wing-covers sprinkled all over with small, irregular black dots. The beetles appear in late summer and feed on the pollen of flowers, ripe fruit and corn in the milk. They go into hibernation and very early the next spring may be seen flying close to the ground with a loud, buzzing sound. The female deposits her white, nearly spherical eggs in the vicinity of manure heaps, in piles of rotting sod and other decaying vegetable matter. When full-grown the larva is some-what over an inch in length, strongly curved and dirty white in color; the posterior part of the body has a dull leaden hue from the contents of the alimentary canal. It differs from the white grub (*Lachnosterna*) in its shorter and more robust form, in the shorter legs and smaller head, and in its habit of crawling upon its back. In July the larvæ pupate within earthen cocoons of a somewhat angular external form. The beetles emerge during August and September. There is only one generation a year.

Hand-picking of the beetles is apparently the most practicable means of controlling this insect when it is found working on ripe fruit or on green corn.

REFERENCES

Slingerland, Can. Ent. XXIX, pp. 50–52. 1897.
U. S. Bur. Ent. Bull. 19, pp. 67–74. 1899.

PEACH STOP-BACK

From Missouri to Alabama, Virginia and northward, nursery-men often experience serious losses from an obscure trouble with peach nursery stock, commonly known as stop-back or peach-sting. When the budded trees are 18 inches to 2 feet in height, the terminal bud turns brown, withers and dies. The stopping of growth of the main branch forces the develop-

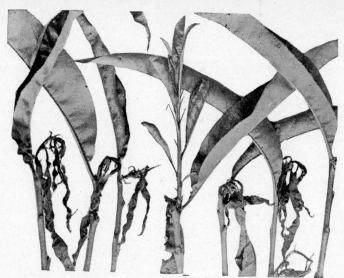

Fig. 255. — Tips of peach nursery trees stung by the tarnished plant-bug.

Fig. 256. — A nearer view of injured peach tips,

300

ment of the laterals, which in turn have their tips killed in a similar way; the result is that the tree takes on a bushy form (Fig. 257) instead of developing a long, straight upright.

Such trees cannot be sold as first-class stock. The loss from this trouble in one New York nursery, amounted in a single year to $15,000.

The cause of peach stop-back has been attributed to a species of thrips (*Euthrips tritici* Fitch), to the peach bud-mite (*Tarsonemus waitei* Banks) and to the tarnished plant-bug (*Lygus pratensis* Linnæus). It is quite certain that in New York, Virginia and Missouri, at least, the latter is responsible for the injury.

The tarnished plant-bug is discussed on page 375. In New York the adults emerge from hibernation in March or April and pass through at least one generation on various weeds before attacking the peach. The last of June or first of July the adults migrate into the blocks of nursery stock, where they puncture the terminal buds and injure them, as described above (Figs. 255 and 256). While

FIG. 257. — A peach nursery tree on which the terminal buds have been killed by the tarnished plant-bug, a condition known as stop-back. Leonard photo.

a few eggs are deposited in the tender tips or in the leaf petioles, very few nymphs are found on peach, practically all of the injury being done by the adults. After three or four weeks the bugs leave the peach, going to wild carrot, wild aster and other weeds. They remain on apple stock for a much longer period.

The control of the tarnished plant-bug on nursery stock is still an unsolved problem. The fact that the injury is caused by the winged adults, and not by the nymphs, makes it impracticable to kill them with a contact spray. They are very shy and active insects that take flight at the slightest alarm. During the daytime they leave the trees several feet in advance of the spray. The terminal shoots of the nursery trees might be protected by inclosing them in paper bags.

References

U. S. Bur. Ent. Bull. 97, Pt. VI. 1912.
Back and Price, Jour. Ec. Ent. V, pp. 329–334. 1912.
Haseman, Jour. Ec. Ent. VI, pp. 237–240. 1913.

Other Peach Insects

GREEN FRUIT-WORMS : *apple*, p. 39.
BUD-MOTH : *apple*, p. 42.
OBLIQUE-BANDED LEAF-ROLLER : *apple*, p. 65.
LEAF-CRUMPLER : *apple*, p. 68.
APPLE-TREE TENT-CATERPILLAR : *apple*, p. 112.
FOREST TENT-CATERPILLAR : *apple*, p. 119.
CLIMBING CUTWORMS : *apple*, p. 138.
SAN JOSÉ SCALE : *apple*, p. 162.
SCURFY SCALE : *apple*, p. 176.
FLAT-HEADED APPLE-TREE BORER : *apple*, p. 194.
TWIG-PRUNER : *apple*, p. 200.
TWIG-GIRDLER : *apple*, p. 202.
FLEA-BEETLES : *apple*, p. 203.
CLOVER-MITE : *apple*, p. 206.
EYE-SPOTTED APPLE-TWIG BORER : *apple*, p. 209.
NEW YORK WEEVIL : *apple*, p. 210.
PEAR THRIPS : *pear*, p. 223.
HOWARD SCALE : *pear*, p. 234.
EUROPEAN PEAR SCALE : *pear*, p. 234.
PLUM CURCULIO : *plum*, p. 243.

CHAPTER IX

CHERRY INSECTS

THE insects of the greatest commercial importance to the cherry grower are the plum curculio (p. 243), the cherry fruit flies, the cherry aphis, the pear slug (p. 214) and on the sweet cherries the San José scale. Cherry trees are also subject to the attack of the fruit-tree bark-beetle (p. 277).

THE CHERRY FRUIT-FLIES

Rhagoletis cingulata Loew and *R. fausta* Osten Sacken

Most wormy cherries in the United States and Canada are caused by the grub of the plum curculio (p. 243), but through-

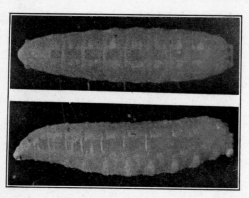

out the northern United States and Canada there occur two closely related species of fruit-flies whose maggots sometimes infest from one fourth to two thirds of the ripening fruits. Unfortunately, there is little external evidence of the work of these

FIG. 258. — Dorsal and side view of the cherry fruit-fly maggot, *R. cingulata* ($\times 7\frac{1}{2}$).

fruit-flies in cherries at picking time, and often the fairest-looking fruits contain the maggots which the housewife may

discover at canning time, or in the bottom of a dish of luscious cherries left over from a previous meal.

FIG. 259. — Cherries infested with fruit-fly maggots.

The species which first attracted attention by its ravages was *R. cingulata*. The full-grown light yellowish-white maggot of this species is about ¼ of an inch in length and scarcely distinguishable from the apple maggot (Fig. 258). There is but a single generation of this cherry fruit-fly annually, the maggots working in the cherries mostly during June, but some may be

FIG. 260. —Cherry cut open and showing a maggot near the pit.

FIG. 261. — Puparia of the cherry fruit-fly (× 9).

found even in August. The infested cherries do not drop, but finally a rotting and sinking in of a portion of the fruit results from the work of the maggots (Fig. 259). When full-grown the maggots leave the cherries, go into the ground about an inch and change to brownish puparia (Fig. 261) which hibernate. The adult insects or fruit-flies begin to emerge about the middle

x

of June and may be found on the trees during the next two months. They are pretty little flies, considerably smaller than a house-fly, and have their wings crossed by four blackish bands (Fig. 262). The body is black, with lateral borders of

thorax light yellow, and the caudal borders of the abdominal segments whitish; the head and legs are yellowish-brown. These flies stick their dirty yellow, elongate eggs through the skin of the ripening cherries, and the maggots, which doubtless hatch in

FIG. 262. — The cherry fruit-fly, *R. cingulata* ($\times 5\frac{1}{2}$).

a few days, revel in the juicy flesh for 3 or 4 weeks, soon forming a rotting cavity near the pit. Rarely does more than one maggot infest the same cherry (Fig. 260). Many of the maggots are nearly full-grown about picking-time and go to the consumer or cannery. Many emerge from the fruits and change to puparia in the bottom of the baskets, and may be carried in this way to new localities.

This fruit-fly may attack many varieties of cherries, whether sweet or sour, early or late.

FIG. 263. — The cherry fruit-fly, *R. fausta* ($\times 5\frac{1}{2}$).

The Morello and Montmorency are varieties often the worst infested. The insect may also work in plums and prunes.

Quite recently it has been discovered that a second species, *R. fausta*, is responsible for a large part of the injury formerly attributed to *cingulata*. It apparently has a somewhat more northern range than that species. In their habits, life history and character of the injury inflicted, the two species are very similar. In New York the flies of *R. fausta* (Fig. 263) appear in early June, but do not begin egg-laying until about two weeks later. During this period they may be observed resting on the foliage and lapping up drops of moisture or feeding on the honey-dew secreted by the cherry aphis. The maggots of *R. fausta* are similar in shape and size to those of *R. cingulata*, but are distinctly more yellow in color; the puparia, moreover, are of a darker brown color than in that species.

Control.

Experiments in New York have shown that the injury from the cherry fruit-flies may be controlled by sprinkling the foliage with sweetened arsenate of lead at the first appearance of the flies in early June. Satisfactory results have been obtained by using arsenate of lead, 5 pounds in 100 gallons of water, sweetened with 3 gallons of cheap molasses. A pint of this mixture is sufficient for a tree of moderate size. In case rains occur, it may be found necessary to make additional applications. The experience of commercial growers also shows that the sweetening of the poisoned spray is unnecessary and that this pest may be controlled by two applications of arsenate of lead, 4 pounds in 100 gallons, made during the time while the flies are emerging.

REFERENCES

Cornell Agr. Exp. Sta. Bull. 172.   1899.
U. S. Bur. Ent. Bull. 44, pp. 70–75.
Cornell Agr. Exp. Sta. Bull. 325.   1912.

THE CHERRY FRUIT-SAWFLY

Hoplocampa cookei Clarke

In Oregon and California young cherries are often infested by the larva of a small blackish sawfly which eats out the kernel of the pit. In one orchard 80 per cent of the fruit was injured in this way.

The adult sawflies appear on the trees in early spring, and the female inserts her smooth, whitish, slightly kidney-shaped egg in one of the sepals or in the upper part of the calyx cup just

before the blossoms open. The eggs hatch in about five days, or just after the petals have fallen. After feeding for a short time on the tissues surrounding the egg cavity, the young larva burrows to the center of the fruit and eats out the kernel; the cherry soon withers and the larva leaves it, only to enter a second or third cherry in which it feeds on the kernel or, after the stone is hardened, on the pulp surrounding it. The larva becomes full-grown in about 24 days, descends to the ground, where at a depth of from 3 to 7 inches it constructs a tough parchment-like cocoon, within which it remains in the larval condition, pupating some time after the winter rains set in. There is only one brood annually.

FIG. 264. — Nest of the cherry-tree tortrix with empty pupa skins protruding.

Control.

No satisfactory means of control have been devised. Thorough cultivation of the soil would doubtless destroy many of the cocoons. Early in the morning the adults are usually sluggish and can be killed with distillate-oil emulsion and tobacco extract as used against the pear thrips. Attempts to kill the larva with arsenate of lead have not given satisfactory results.

REFERENCE

U. S. Bur. Ent. Bull. 116, Pt. III. 1913.

THE CHERRY-TREE TORTRIX

Archips cerasivorana Fitch

In June and July the ends of the branches of both the wild and cultivated cherry are often seen inclosed in large, pointed silken nests (Fig. 264). Within each nest there lives a colony of lemon-yellow larvæ, about ⅝ inch in length when full-grown, all the offspring of a single ochre-yellow moth. As the larvæ increase in size the nest becomes filled with large, dark-colored masses of excrement webbed together with silk (Fig. 265). The larvæ mature early in July and pupate

Fig. 265. — A mass of excrement from interior of nest with pupa skins attached.

Fig. 266. — The cherry-tree tortrix moth (× 2).

inside the nest. When about to transform, in the latter part of July in New York, the pupæ work their way out of the nest, clinging to it only by the hooks at

FIG. 267. — Egg mass of the cherry-tree tortrix with a moth resting on a cherry branch.

the end of the body.   When the moths emerge, empty pupa cases are left projecting from the nest.   The moths (Fig. 266) expand from $\frac{4}{5}$ to $1\frac{1}{5}$ inches; the front wings are bright ochre-yellow, marked with irregular brownish spots and numerous transverse bands of a pale leaden blue.   The eggs are deposited in flattened masses on the smaller branches and are protected by a gluey covering (Figs. 267 and 268).

This insect rarely becomes troublesome on cultivated cherries.   The webs should be cut out and burned.

FIG. 268. — Egg-mass enlarged.

REFERENCE

Cornell Agr. Exp. Sta. Bull. 23, pp. 113–115.   1890.

THE CHERRY PLANT-LOUSE

Myzus cerasi Fabricius

This blackish plant-louse very generally infests the cherry both in Europe, in the United States and Canada, east of the Rocky Mountains; it also occurs in California.   Sweet cherries are more liable to injury than the sour varieties; this is sometimes strikingly shown where both kinds are grown together in nurseries (Fig. 271).   The sweet cherries may have the leaves badly curled and the new growth stunted, while sour cherries growing in the next row show little or no injury.

The shining black winter eggs are found attached to the bark of the smaller branches, mostly around the buds.   They hatch about the time the buds open, and the stem-mothers found colonies of wingless females (Fig. 269) which reproduce with wonderful rapidity, so that within a few weeks the tips of the

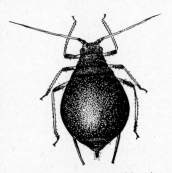

new growth and the under side of the leaves are thickly covered with the lice. The leaves become curled, the tips of the branches

Fig. 269. — The cherry aphis, wingless viviparous female, enlarged.

Fig. 270. — The cherry aphis, winged viviparous female, enlarged.

cease growing and in severe cases the fruit may be attacked and drop. Nursery stock is especially liable to be injured.

Fig. 271. — Sweet cherry nursery trees badly infested with the cherry aphis.

As the feeding quarters become crowded, winged forms (Fig. 270) are produced; whether they merely fly to other cherry-trees or establish summer colonies on other food-plants is unsettled. In Colorado a few lice at least remain on the cherry throughout the season. No alternate food-plant has been found. Sexual forms are produced, and the winter eggs are deposited in September and October.

Treatment.

This insect is not difficult to control by spraying with kerosene emulsion, whale-oil soap solution or tobacco extract, provided the work is done as soon as the lice appear and before the leaves become curled. On nursery stock the lice attack and curl the leaves on the tips of the young trees where it is impossible to hit them with a spray. In such cases it has been found practicable to dip the infested tips in a solution of whale-oil soap, 1 pound in 7 gallons of water. The solution is carried in a pail, and the tips of the young trees are carefully bent over and held in the liquid long enough to wet all the lice.

REFERENCES

Lintner, 5th Rept. N. Y. State Ent., pp. 253–257. 1889.
Ohio Agr. Exp. Sta. Bull. Tech. Ser., Vol. I, No. 2, pp. 111–113. 1890.
N. Y. (Geneva) Agr. Exp. Sta. Bull. 136, p. 598. 1897.
Col. Agr. Exp. Sta. Bull. 133, pp. 42–44. 1908.

THE CHERRY SCALE OR FORBES' SCALE

Aspidiotus forbesi Johnson

Discovered in Illinois in 1896, this scale insect has since been found in widely separated localities throughout the United States. Cherry trees, both wild and cultivated, seem to be its favorite food-plants, the trunks, branches and sometimes the leaves and fruits being attacked. It also occurs on apple, apricot, pear, plum, quince and currant. Externally the mature, nearly circular, dark grayish female scales (Fig. 272) are indistinguishable from Putnam's scale or the European fruit-scale, but the reddish-orange, nearly central exuvial spot gives it quite a different appearance from the San José scale. The cherry scale winters in a half-grown condition in Illinois, and it is apparently both oviparous and ovoviviparous, the young beginning to emerge early in May, and eggs and young occurring as late as the middle of June. Two broods are produced annually in Illinois, the young of the second generation appearing in August and September. Possibly a third brood occurs farther south. Seven

Fig. 272.—The cherry scale, males and females. Redrawn after Joutel.

minute parasites have been bred from this cherry scale, and the twice-stabbed ladybird beetles and their grubs feed upon it.

Thorough applications of the winter washes as recommended for the San José scale readily control this scale.

OTHER CHERRY INSECTS

BUD-MOTH: *apple*, p. 42.
FRUIT-TREE LEAF-ROLLER: *apple*, p. 62.
OBLIQUE-BANDED LEAF-ROLLER: *apple*, p. 65.
LEAF-CRUMPLER: *apple*, p. 68.
CANKER-WORM: *apple*, p. 77.
CALIFORNIA TUSSOCK-MOTH: *apple*, p. 104.
ORIENTAL MOTH: *apple*, p. 106.
FOREST TENT-CATERPILLAR: *apple*, p. 119.
YELLOW-NECKED APPLE CATERPILLAR: *apple*, p. 123.
RED-HUMPED APPLE CATERPILLAR: *apple*, p. 125.
CLIMBING CUTWORMS: *apple*, p. 138.
BUFFALO TREE-HOPPER: *apple*, p. 160.
SAN JOSÉ SCALE: *apple*, p. 162.

CHAPTER X

RASPBERRY, BLACKBERRY AND DEWBERRY INSECTS

THESE fruits are closely related botanically and are in general subject to the attack of the same insects; their enemies are here treated together for the sake of convenience. As a rule the raspberry and blackberry are not so liable to serious attack as other fruits, and spraying is only occasionally necessary.

THE RED-SPIDER

Tetranychus bimaculatus Harvey

The red-spider is a minute mite (Fig. 273), about $\frac{1}{50}$ inch in length, varying in color from pale greenish-yellow to dark crimson-red and usually marked with two dark spots at the side of the body. It has long been known as a greenhouse pest, where it attacks a great variety of plants; it sometimes injures peach, raspberry, currant and rose in the open. It thrives in a warm, dry atmosphere and is most abundant out of doors in seasons of drought.

Red-spiders are usually found on the under side of the leaves, where they live under the protection of a very delicate silken web; in feeding they break the epidermis of the leaf and suck out the sap, causing the foliage to turn yellowish in spots. The female deposits her minute, elongate, transparent eggs on the surface of the leaf. The young mites resemble the adults, but are lighter in color and have only six legs instead of eight. There are several generations a season in the open, while in greenhouses breeding is continuous the year round. The adults hibernate in the ground or beneath suitable shelter and crawl back to the trees in early spring. (See also page 208.)

Treatment.

Red-spiders may be controlled in greenhouses by the use of sulfur, either as a dust or in water, or by persistent spraying with water, taking care to hit the under side of the leaves. Use lots of force and little water, to avoid drenching the beds.

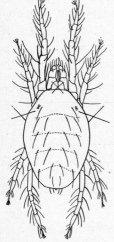

When infesting plants in the open, the mites may be destroyed by dusting with sulfur or by spraying with a mixture of 1 pound of finely powdered sulfur in 3 gallons of water, in which a little soap has been dissolved. The sulfur settles quickly, and the mixture should be agitated constantly during the spraying. It will help to keep the sulfur in suspension if it is first made into a paste with water containing $\frac{1}{2}$ of 1 per cent of glue. The boiled lime-sulfur solution cannot be used on raspberries, as it is likely to cause foliage injury.

Excellent results in the control of red-spider have been reported in California from the use of a flour paste prepared as follows:

Fig. 273. — The red spider. Redrawn after Miss M. A. Palmer (×66).

Mix a cheap grade of wheat flour with cold water, making a thin batter, without lumps; or wash the flour through a wire screen with a stream of cold water. Dilute until there is one pound of flour in each gallon of mixture. Cook until a paste is formed, stirring constantly to prevent caking or burning. Add sufficient water to make up for evaporation. For use, add 8 gallons of this stock solution to 100 gallons of water. When mixed in the spray tank flour paste has a tendency to settle and in order to do satisfactory work agitation is necessary.

REFERENCES

Me. Agr. Exp. Sta. Rept. for 1892, pp. 133–146.   1893.

Col. Agr. Exp. Sta. Bull. 152.   1909.

U. S. Bur. Ent. Circ. 166.   1913.

THE BLACKBERRY LEAF-MINER

Metallus rubi Forbes

The leaves of blackberries and dewberries are often injured to a considerable extent in the Eastern states and Canada from Delaware and Missouri northward by the larva of a small, nearly black sawfly, about ⅙ inch in length (Fig. 274). The larvæ feed between the two layers of the leaf, excavating a rather large, irregular blotched mine; three or four mines may occur in a single leaf (Fig. 276).

FIG. 274. — The blackberry leaf-miner, adult.   Enlarged.

The injured portion of the leaf turns brown and dies; in severe cases the whole field has the appearance of having been singed by fire.

The adults appear in late May or early June, and the female inserts her white, flattened egg into the tissue of the leaf through a puncture in the upper surface (Fig. 277). The egg lies next to the lower epidermis beneath a low blister about $\frac{1}{30}$ inch in diameter. At the time of hatching the larva has a very large head in proportion to its size. When full-grown it is about ⅓ inch in length, and greenish-white in color with brownish markings (Fig. 275).

FIG. 275. — Full-grown larvæ of the blackberry leaf-miner ($\times 6\frac{3}{4}$).

In New York the majority of the larvæ of the first brood become full-grown in July; in Delaware they are nearly a month earlier. They go into the ground an inch or so to transform, the second brood adults appearing in August in New York. There are two apparently full broods in Delaware; in New

FIG. 276. — Mines in a blackberry leaf.

York the second brood is probably only partial. The winter is passed in the larval or pupal condition in the ground.

A satisfactory method of controlling this insect has not been worked out, but it would be worth while to try to kill the larvæ in their burrows by spraying with " Black Leaf 40" tobacco extract, 1 pint in 100 gallons of water, to which a little soap has been added, 2 pounds in 50 gallons. This mixture has been found of value against a similar larva mining the leaves of the European elm.

Fig. 277. — Egg-blisters of the blackberry leaf-miner, one opened to show the egg (×6).

REFERENCE

Del. Agr. Exp. Sta. Bull. 87, pp. 10–15. 1910.

THE RASPBERRY SAWFLY

Monophadnus rubi Harris

In the Northern states and Canada from Iowa eastward the larva of this sawfly often causes serious injury to the raspberry and also attacks the blackberry and dewberry to a slight extent. The adult is a black, thick-bodied, four-winged fly about ¼ inch in length ; the female has a broad, yellowish-white band across the abdomen. The flies appear in May, and the female deposits her eggs singly between the two layers of the leaf near a prominent vein. They are placed in position through a small incision in the lower epidermis of the leaf made by the saw-like ovipositor of the insect. When first laid, the egg is white, long, oval in form, obtusely rounded at the ends, and is about $\frac{1}{30}$ inch in length; before hatching it becomes nearly pear-shaped and increases in length to over $\frac{1}{20}$ inch. The leaf tissue above the egg becomes dry and somewhat withered and finally turns to a light yellow color, giving the leaf a characteristic spotted appearance.

The eggs hatch in a week or ten days. At first the larvæ feed on the outer epidermis only, but as they grow larger eat out irregular holes through the leaf, and finally leave only the larger veins. When sufficiently abundant to devour all the foliage, they often attack the tender bark of the new growth. They also feed to a slight extent on the blossom buds and immature fruits.

The full-grown larvæ are about $\frac{3}{4}$ inch in length, light green in color and covered with spine-bearing tubercles arranged in double transverse rows. They become mature in about ten days, and then crawl to the ground, where at a depth of two or three inches they construct oblong, nearly cylindrical cocoons composed of a dark brown mucilaginous substance mixed with strands of silk and particles of earth. The larva remains within the cocoon in a quiescent condition, known as the prepupa, until the following May, when it transforms to a pale green pupa, and the adults emerge in a few days.

Treatment.

If the soil is well cultivated and free from weeds, the larvæ may be brushed from the bushes during the heat of the day and will perish without being able to regain their food-plant. Pine branches are often used for this purpose. The larvæ are easily killed by arsenical sprays, but as there is some danger in their use on ripening fruit, it is better to use hellebore, 1 ounce in 1 gallon of water.

REFERENCE

N. Y. (Geneva) Agr. Exp. Sta. Bull. 150. 1898.

The Raspberry Webworm

Pamphilius fletcheri MacGillivray

This insect has become troublesome only in New Brunswick, and little is known of its life history. The smooth, bright green larvæ, half an inch in length when full-grown, web together the terminal leaves of the raspberry, making a tent within which they feed. The adult sawfly is a little less than $\frac{3}{8}$ inch in length; the head and thorax are black, marked with white; in the female the front third of the abdomen is black and the rest reddish-yellow; in the male the abdomen is black with a broad yellowish band across the middle. The adults appear about the middle of June, and the larvæ develop during the next few weeks. The eggs are unknown. The winter is probably passed by the larvæ in the ground.

This rather uncommon pest can be controlled by persistent hand-picking of infested leaves or by dusting the plants with hellebore soon after the larvæ hatch, before they have webbed the leaves together.

REFERENCE

Fletcher, Rept. Ent. Bot. for 1899, p. 180. 1900.

The Raspberry Leaf-roller

Exartema permundanum Clemens

The terminal leaves of the raspberry are sometimes webbed together in May and early June into a more or less twisted mass by a small dark green larva with a pitchy-black head and thoracic shield. This caterpillar is sometimes destructive to strawberries by webbing together the clusters of flowers and flower buds. When full-grown the larva usually folds over a part of a leaf, forming a cavity within which it pupates. The moths appear in about two weeks; they have a wing expanse

Y

of about one half inch; the fore wings are dull yellowish or greenish-brown with irregular lighter markings crossing the wing obliquely; the hind wings are ashy brown. The eggs are unknown. There are probably two broods annually, the winter being passed in the egg state. While generally distributed over the Eastern states, this insect rarely causes noticeable damage; it also feeds on the wild blackberry, hazel, meadow sweet and hickory.

In small plantings this insect can be controlled by hand-picking the infested tips and crushing the larvæ. On a larger scale it would be more economical to use some arsenical spray when the larvæ first appear in the spring.

REFERENCES

Comstock, Rept. U. S. Com. Agr., 1880, p. 267.
Packard, 5th Rept. U. S. Ent. Com., p. 312. 1890.
Ohio Agr. Exp. Sta. Bull. 45, p. 181. 1893.

THE BLACKBERRY PSYLLID

Trioza tripunctata Fitch

This jumping plant-louse, a near relative of the pear psylla, occurs in the Atlantic states from Virginia to Maine. It has been recorded as injuring the cultivated blackberry in New Jersey and Maine, and on Long Island. Its native food-plant is the wild blackberry. Its reported occurrence on pine is doubtless purely accidental.

The adult insect is about $\frac{1}{6}$ inch in length; the body is yellowish-brown, the eyes dark brown and the wings marked by three yellowish-brown bands. The insect hibernates as an adult. The flies appear on the blackberry soon after growth starts in the spring and deposit their minute, light yellow eggs in the pubescence of the leaf petioles and young canes. On Long Island adults, eggs and newly hatched nymphs were observed the latter part of June. Both adults and nymphs puncture the leaves and tender canes with their piercing mouth parts and feed on the juices of the plant, causing the leaves to curl, also dwarfing and distorting the young canes. The minute, young nymphs are whitish or greenish-white in color; the older nymphs are yellowish. They mature in early fall, and the adults go into hibernation.

Little experimental work has been done in the control of this insect, but it is probable that some of the measures employed against the pear psylla would be found satisfactory.

REFERENCE

N. Y. (Geneva) Agr. Exp. Sta. 14th Rept. for 1895, pp. 619–623. 1896.

THE AMERICAN RASPBERRY BEETLE

Byturus unicolor Say

The red raspberry crop is sometimes severely injured locally in the eastern United States and Canada by this small, slightly hairy, light brown beetle, about $\frac{1}{5}$ inch in length. The beetles appear in May and begin feeding on the buds and tender leaves, and later attack the blossoms themselves. When the cluster of blossom buds is disclosed, they work their way in between them and eat holes into the buds. They also skeletonize the leaves to a considerable extent and feed on the stamens and pistils of the blossoms, often causing an almost total failure of the crop.

Although the egg is unknown, it is probably deposited on the fruit, for the young grubs are found burrowing through the receptacle or lying upon its surface beneath the berry. When full-grown they are nearly white in color and about $\frac{1}{4}$ inch in length. When abundant many of the grubs adhere to the berry at picking time, making it necessary to carefully hand-pick the fruit intended for table use. Normally they fall to the earth and hibernate as pupæ under trash or in the soil near the surface.

Treatment.

Experiments conducted in Ohio have shown that this pest can be controlled by a thorough application of arsenate of lead, 6 to 8 pounds to 100 gallons of water, made as the first beetles appear. Thorough shallow cultivation of the soil in the fall would doubtless destroy many of the pupæ in their hibernating quarters.

REFERENCES

Fitch, Trans. N. Y. State Agr. Soc. for 1870, pp. 358–360. 1872.
Felt, 14th Rept. N. Y. State Ent., pp. 158–160. 1898.
Ohio Agr. Exp. Sta. Bull. 202. 1909.

THE NEGRO-BUG

Corimelæna pulicaria Germar

FIG. 278. — A tree-cricket ovipositing in a raspberry cane.

Blackberries, raspberries and strawberries often acquire an unpleasant flavor from having served as the feeding ground of a small black sucking bug. The adult is shiny black and has a white stripe on each side of the body; it is about an eighth of an inch in length. The female deposits her orange-yellow, elongate, oval eggs singly on the leaves of the plant. They hatch in about sixteen days. The nymphs puncture the tender foliage as well as the fruit, sometimes causing a slight injury to the leaves. On berries, however, they are most troublesome because of the disgusting odor which they impart to the fruit. This insect is sometimes a serious celery pest.

No satisfactory method of controlling negro bugs on berries has yet been devised.

The Tree-cricket

Œcanthus nigricornis Walker

Several species of tree-crickets occur abundantly in the eastern United States and Canada. These delicate, greenish-

Fig. 279. — Tree-cricket egg-scars in raspberry canes, one cane split open to show the eggs.

white, long-horned crickets become mature in late summer and the females deposit their eggs in punctures in the tissue of various plants. Injury to blackberry and raspberry canes is caused principally by the female of *nigricornis*. The eggs of this species are about $\frac{1}{8}$ inch in length, cylindrical, slightly curved and chrome yellow in color, with the egg-cap cream-colored. They are inserted (Fig. 278) in a row of punctures often two inches in length, each row on the average containing about 30 eggs and may have as many as 80 in a row (Figs. 279 and 280). The eggs hatch in May and June and the young tree crickets feed principally on aphids and other soft-bodied insects.

The rows of punctures either kill the upper part of the cane or so weaken it as to prevent the development of the fruit. When very abundant, as is sometimes the case, the loss may be large.

In the past there has been some confusion as to the identity of the species ovipositing in raspberry canes. This injury was

Fig. 280. — Tree-cricket eggs enlarged.

formerly attributed to *O. niveus*, but recent work at the Geneva Experiment Station has shown that *O. nigricornis* is the real culprit and that *O. niveus* deposits its eggs preferably in the bark of the smaller branches of apple and other trees (see p. 211).

The tree-crickets injuring berry canes can be held in check by systematically collecting the canes containing the eggs at the time of pruning and destroying them.

Reference

Parrott, Jour. Ec. Ent. IV, pp. 216–218, pl. 6. 1911.

The Raspberry Cane-borer

Oberea bimaculata Olivier

This native American borer often causes considerable injury to the black and red raspberry and to the blackberry; its original food-plant was the wild raspberry. It is generally distributed over the Northern states and Canada. The adult is a long-horned, slender-bodied beetle about half an inch in length (Fig. 284). It is of a deep black color except the prothorax, which is yellow, usually with two or three black spots on the upper surface. The beetles appear in June and the female deposits her eggs (Fig. 281) singly in the pith of the tender

FIG. 281. — The raspberry cane-borer girdling a cane after oviposition.

new growth about six inches from the tip of the cane. She first makes two rows of punctures encircling the cane about half an inch apart, and between them, but nearer the lower row, inserts the egg in a deep puncture directed upward (Figs. 282 and 285). Sometimes the beetle girdles the cane spirally, and abandons the cane without ovipositing. The scars (Fig. 283) of these imperfect girdles are common on blackberry. The egg is yellowish-white, elongate, nearly cylindrical, with rounded ends slightly curved, and is about $\frac{1}{10}$ inch in length. The

FIG. 282. — A raspberry cane girdled above and below the egg-puncture of the cane-borer.

FIG. 283. — Scar of an imperfect girdle of the raspberry cane-borer; common on blackberry.

girdling of the cane causes the tip to wilt, and is supposed to protect the egg from being crushed by the rapidly growing tissue in which it lies.

The eggs hatch in early July and the borer eats out a burrow towards the base of the cane, passing the winter in a partly grown condition only an inch or two below the girdle

FIG. 284. — The raspberry cane-borer beetle (× 2⅔). Knight photo.

FIG. 285. — Egg of the raspberry cane-borer in position (× 3¾).

(Fig. 286). The second season it continues its burrow through what is now the bearing cane and usually kills it before the fruit matures. It reaches the base of the cane by fall and there passes the second winter in its burrow below the surface of the ground. The burrow winds through the pith and at frequent intervals an opening is made in the bark through which the larva casts forth its excrement. The larva pupates in its burrow the second spring after the laying of the egg and the beetle emerges in May and June.

Treatment.

The tips of the young canes in which the eggs are deposited soon wilt and are easily seen. As soon as noticed they should be cut off below the lower girdle and destroyed. Likewise when bearing canes are found infested with the borers they should be cut close to the ground and burned. This pest would probably be much more abundant were it not the common practice to cut out and burn all old canes after the crop has been picked, thus destroying all the nearly mature borers.

Fig. 286. — Raspberry cane-borer hibernating in a short burrow at the tip of a cane.

REFERENCES

Cornell Agr. Exp. Sta. Bull. 23, pp. 122–124. 1890.
Ohio Agr. Exp. Sta. Bull. 96, pp. 20–22. 1898.

THE RASPBERRY CANE-MAGGOT

Phorbia rubivora Coquillett

Black and red raspberries and blackberries are sometimes severely injured in the Northern states and Canada by a small white maggot which burrows in the new canes and kills them. It also occurs in western Washington. The parent fly (Fig. 287) closely resembles the house fly, but is somewhat smaller.

Fig. 287. — Adult of the raspberry cane-maggot. Enlarged.

The flies appear in the latter part of April, when the new raspberry shoots are a few inches in height, and the female deposits her comparatively large, prettily sculptured, elongate white eggs loosely near the tip of the shoot in the crotch formed by the bases of the tip leaves (Fig. 288). How soon the eggs hatch is unknown, but it is doubtless in a few days.

On leaving the egg shell the young maggot crawls down the shoot for a short distance and then burrows its way into the pith. After tunneling about half the length of the shoot the maggot works its way nearly out to the bark and deftly continues its tunnel around the shoot, thus girdling it from the inside (Fig. 289). The part of the shoot above the girdle soon wilts, shrinks in size and droops over. Soon after the tip droops, a

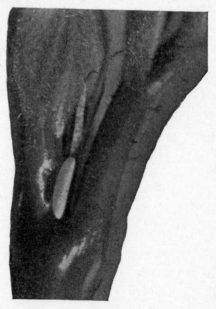

Fig. 288. — Egg of the raspberry cane-maggot in position at the base of a leaf (× 8).

dry rot sets in at the girdled point and the whole shoot usually dries up and dies.

After checking the growth in May the maggot proceeds to burrow downward in the pith, usually reaching the base near the surface of the ground some time in June. In late June and early July the maggots, without leaving their burrows, transform to pupæ within the hardened dark brown larval skin or puparium. The adult, however, does not emerge till the following April, there being but one brood a year.

FIG. 289. — Raspberry shoots girdled by the cane-maggots.

Control.

With a little watchfulness this raspberry pest can be easily checked. Its presence can be quickly detected in May, as its work is then very conspicuous. The remedy is simple. As soon as a drooping tip is seen, either pull up the shoot or cut it off several inches below the girdle and burn it. This method faithfully carried out throughout May will quickly check the pest. There is no possible chance of getting at the insect with a spray. Simply burn all infested shoots in May.

REFERENCES

Cornell Agr. Exp. Sta. Bull. 126, pp. 54–60. 1897.
Wash. Agr. Exp. Sta. Bull. 62. 1904.

The Red-necked Cane-borer

Agrilus ruficollis Fabricius

The new canes of blackberry, dewberry and raspberry are often injured by the larva of a beetle which causes irregular swellings or galls from one to three inches in length. These galls are gradual enlargements of the cane and are characterized by a longitudinal splitting of the bark (Fig. 293). Infested canes either die or are so weakened as to prevent the development of the fruit.

Fig. 290. — Beetle of the red-necked cane-borer ($\times 2\frac{5}{8}$).

Fig. 291. — A beetle that died because she was unable to free her ovipositor after egg-laying. Knight photo.

The parent beetle (Fig. 290) is about $\frac{1}{3}$ inch in length; the wing-covers are black with a dull bluish reflection, and the thorax or "neck" is metallic coppery, reddish or brassy; the head is black with metallic reflections. The beetles may be found on the berry bushes on bright days from the last of May till August, but are most abundant in June. The egg is inserted in the bark near the base of a leaf on the new growth. Figure 291 shows a female beetle that died because she was unable to free her ovipositor after depositing the egg. The young larva burrows upward in the sapwood, passing around the stem in a spiral course from two to six times, thus girdling the cane and causing the gall (Fig. 293). The larva then bores into the pith and usually burrows upwards, hibernating in the pith several inches from the point of entrance. It is then a yellowish-white, slender, somewhat flattened grub, about $\frac{5}{8}$

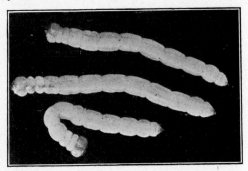

Fig. 292. — Full-grown red-necked cane-borer larvæ (× 3).

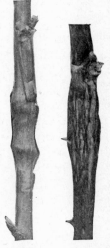

to $\frac{3}{4}$ inch in length, with a brownish head (Fig. 292); the tip of the abdomen is provided with two brown hooks. Early in the spring it completes its growth and transforms into a whitish pupa in an oval cell in the pith. The beetles emerge from May till July.

Fig. 293. — A red raspberry cane showing the spiral course of the burrow; the more normal form of the gall on blackberry.

This pest may be controlled by cutting out and burning all infested canes during the fall, winter or early spring. This work can be done best while pruning the bushes. All wild or neglected berry bushes in which the beetles may breed should be destroyed.

REFERENCES

Smith, Ins. Life, IV, pp. 27–30. 1891.
N. J. Agr. Exp. Sta. Spec. Bull. N., pp. 4–8. 1891.
W. Va. Agr. Exp. Sta. Bull. 15. 1891.
Ohio Agr. Exp. Sta. Bull. 44, pp. 191–193. 1893.

- ## THE RASPBERRY HORNTAIL
Hartigia abdominalis Cresson

In California the tender tips of the young shoots of the raspberry, blackberry, loganberry and rose are often girdled and killed by the spiral burrows of the young larvæ of a small yellow and black horntail fly. The adults are slender, four-winged flies about $\frac{5}{8}$ of an inch in length. They appear on the berry bushes the last of April and remain abundant until in August. The female inserts her smooth, pearly-white, flattened oval eggs, about $\frac{1}{16}$ inch in length, singly just under the bark of the young canes. The newly-hatched, yellowish-brown larvæ feed for a time in the vicinity of the egg, and when about $\frac{1}{4}$ of an inch in length burrow spirally downward three or four times around the cane, working just beneath the bark and thus girdling the tip. The larva then enters the pith and burrows upward until the tip of the branch dies, when it turns round in its burrow and works down through the pith, towards the base of the cane. The larvæ become full-grown in from four to six months and are then nearly an inch in length and nearly white in color. They pupate at the end of the burrow and the adults gnaw their way out of the cane. There are said to be several broods annually.

The location of the egg may be easily determined by the discoloration of the surrounding tissue, and it may easily be crushed by hand. This is probably the most feasible method of controlling the pest. Many of the larvæ might also be destroyed by cutting off the dying tips of the canes as soon as wilting of the leaves is observed. This work cannot be done in the winter as it is then difficult to distinguish the infested canes.

REFERENCE

Monthly Bull. St. Com. Hort. Cal., I, No. 12, pp. 889–901. 1912.

The Blackberry Crown-borer

Bembecia marginata Harris

Blackberries and raspberries are often injured by the caterpillar of a clear-wing moth which burrows in the roots and crown. This insect is generally distributed throughout the Northern states and Canada and has been recorded from New Mexico.

The parent moth has an expanse of 1 to $1\frac{1}{4}$ inches; the wings are transparent, with a brown margin, and the fore wings have a narrow, transverse brown band on the outer third; the abdomen is black crossed by four bands of bright yellow; in the female the last segment is yellow, in the male, black mixed with yellow. The moths emerge during August and early September and the female deposits her oval, reddish-brown eggs, $\frac{1}{16}$ inch in length, singly on the under side of the leaves near the edge. Each female lays about 140 eggs.

On hatching, the caterpillar is about $\frac{1}{16}$ inch in length and is nearly white, with a brownish head. It crawls down the stem and goes into hibernation curled up in a small cavity beneath a blister-like elevation of the bark of the cane just below

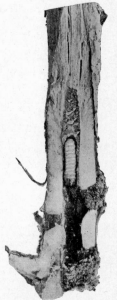

Fig. 294. — The blackberry crown-borer in its burrow at the base of a plant.

the surface of the ground or may hibernate in crevices at the base of the canes or under flakes of bark. In the spring the caterpillars enter the roots or the base of the cane where they generally burrow just beneath the bark, girdling that part of the plant. By the second winter the larvæ are $\frac{1}{2}$ to $\frac{3}{4}$ inch in length. They hibernate in their burrows, and the following spring take an upward course, either through the pith or in the wood just beneath the bark (Fig. 294). They become full-grown (Fig. 295) in July and burrow out to the surface of the cane but leave the epidermis intact over the opening. The pupa is about ¾ inch in length, of a reddish-brown color, and has the front end armed with a sharp-pointed process used in breaking away the epidermis over the end of the burrow. The pupal stage lasts 25 to 30 days. When about to transform the pupa works itself part way out of the burrow, so that after the moth has emerged the empty pupal skin is left protruding

from the opening. The moths usually emerge in the afternoon and mating takes place in the early evening.

This borer may be held in check by systematically digging out the larvæ

Fig. 295. — Mature larva of the blackberry crown-borer (× 2).

whenever a dying or wilting tip indicates its presence. All wild berry bushes in the vicinity of berry fields should be destroyed to prevent the breeding of the moths.

References

N. J. Agr. Exp. Sta. Spec. Bull. N., pp. 9–12. 1891.
Engel, Ent. News, XV, pp. 68–71. 1904.
Wash. Agr. Exp. Sta. Bull. 63. 1904.

The Rose Scale

Aulacaspis (Diaspis) rosæ Bouché

The stems of roses, blackberry, raspberry and dewberry growing in damp, shady places often become densely coated with a snow-white, nearly circular scale-insect, the larger ones about 1/10 of an inch in diameter, with the two light yellow exuviæ or cast skins at the margin. Among these larger female scales are many of the shorter, narrower, three-ridged, white scales of the male insect (Fig. 297).

This rose scale is practically a cosmopolitan insect, occurring wherever roses are grown, and it is widely distributed over the United States and Canada. It is not often a serious pest in berry plantations and is usually readily controlled.

In New Jersey, and doubtless also in more southern localities, the rose scale may hibernate in all stages from the egg to the gravid females, mostly, however, as young scales of both sexes, as male pupæ and as full-grown females. Observations in Canada also indicate similar hibernation conditions and at least two generations annually. Breeding is almost continuous after April, and three or more broods may occur in New Jersey and southward.

Fig. 296. — The rose scale, tipped over to show the eggs beneath it ($\times 7\frac{1}{2}$).

Fig. 297. — A male and two female rose scales ($\times 3$).

Two little hymenopterous parasites, *Aphelinus diaspidis* and *Arrhenophagus chionaspidis*, destroy many of the scales.

Remedial measures.

In berry plantations cut and burn all badly infested canes soon after the fruit is off, or in winter, and thus prevent serious infestation of the new canes. Thorough applications of a soap spray (1 pound whale-oil or other good soap in 1 gallon water) or the lime-sulfur spray in winter or early spring have been found to effectively control this insect.

z

Other Raspberry and Blackberry Insects

Bud-moth : *apple*, p. 42.
Fruit-tree leaf-roller : *apple*, p. 62.
Oblique-banded leaf-roller : *apple*, p. 65.
Half-winged geometer : *apple*, p. 96.
Red-humped apple caterpillar : *apple*, p. 125.
Climbing cutworms : *apple*, p. 138.
Oyster-shell scale : *apple*, p. 171.
Scurfy scale : *apple*, p. 176.
Apple leaf-hopper : *apple*, p. 180.
Flea-beetles : *apple*, p. 203.
Clover-mite : *apple*, p. 206.
European fruit lecanium : *plum*, p. 261.
Rose chafer : *grape*, p. 397.
Imbricated snout-beetle : *strawberry*, p. 371.
Fuller's rose beetle : *strawberry*, p. 389.
Strawberry root-worms : *strawberry*, p. 391.

CHAPTER XI

CURRANT AND GOOSEBERRY INSECTS

COMMERCIALLY the most important insect pests of the currant and gooseberry are the imported currant worm, the San José scale and the currant plant-louse.

THE IMPORTED CURRANT BORER

Sesia tipuliformis Clerck

This destructive European currant-borer was introduced into this country some time before 1826, and is now widely distributed throughout North America; it also occurs in Asia, Australia and New Zealand. The caterpillars burrow (Fig. 298) in the smaller canes and eat out the pith for a distance of several inches, causing the branch to die. While not often a serious pest this insect has been known to become so abundant in certain gardens that the raising of currants was abandoned for a time.

The beautiful, clear-winged moths (Fig. 300) appear in June and may be found flying rapidly about the plants or resting on the leaves. The female moth has an expanse of about $\frac{3}{4}$ inch; the wings are transparent, with a border of golden purple and a bar of the same color across the fore wing; the body is purplish black with three narrow bands of yellow on the abdomen in the female and four in the male.

The female moth deposits her brown, almost globular eggs singly on the bark. The young larvæ bore into the stem and then burrow up or down, through the pith, killing the cane.

They become nearly full-grown by fall and hibernate in the burrow. The larvæ are then about ½ inch in length, whitish

FIG. 299. — Pupa of the imported currant borer. Matheson photo (× 4).

FIG. 298. — The imported currant borer in its burrow ready to pupate (× 1½).

FIG. 300. — Moth of the imported currant borer.

with a brownish head and legs. In May the larva burrows out to the surface of the stem, leaving the opening covered by a thin layer of bark. It then transforms to a pupa (Fig. 299) in a silken-lined cavity at the end of the burrow. When ready to transform the pupa pushes itself part way out of the opening and the moth leaves the empty pupal skin projecting from the cavity. There is only one brood a year.

The infested canes do not die in the fall but are usually able to put out a sickly foliage the following spring. Such canes should be cut off and destroyed before June 1 to prevent the emergence of the moths.

REFERENCES

Col. Agr. Exp. Sta. Bull. 19, pp. 21–22. 1892.
Vt. Agr. Exp. Sta. Rept. for 1894, pp. 130–132. 1895.
Wash. Agr. Exp. Sta. Bull. 36, p. 14. 1898.

The Imported Currant Worm

Pteronus ribesii Scopoli

Introduced into this country from Europe about 1857, this insect now occurs wherever currants or gooseberries are grown in the northeastern states and Canada, and is the commonest and best known of garden pests.

The adult sawflies (Fig. 301) appear as soon as the leaves put forth in the spring; the female is about one third inch in length with the head and thorax more or less black when viewed from above and has the abdomen dull yellowish; the male is smaller,

Fig. 301. — Adults of the imported currant worm (× 2).

FIG. 302. — Eggs of the imported currant worm.

one fourth inch in length, and has the dorsal aspect of the abdomen black except at tip. As soon as the leaves have expanded the female deposits her whitish, smooth, elongate eggs end to end in rows along the principal veins on the underside of the leaves of the food-plant (Fig. 302).

The eggs increase in size considerably after deposition, and measure about $\frac{1}{20}$ inch in length when ready to hatch. They hatch in a week or ten days, and the small whitish larvæ begin their destructive work by eating small holes through the leaves (Fig. 303). As they increase in size the color changes to green, and after the first molt the body becomes covered with many black spots and the head is black.

While small the larvæ feed in colonies, 30 or 40 on a leaf, which is soon consumed; they then scatter

FIG. 303. — Newly hatched currant worms feeding.

to other parts of the plant. With increase in size they become more destructive, and if numerous are able to strip a bush of its leaves in a few days (Fig. 304). The larvæ molt several times as they increase in size, but the exact number of molts has not been determined. They become full-grown in two or three weeks, and are then about ¾ inch in length. At the

last molt they lose their black spots and assume a uniform green color tinged with yellow at the ends. The larva then descends to the ground and spins a smooth oval brownish cocoon beneath leaves or other trash; sometimes the cocoons are attached to the

FIG. 304. — Full-grown currant worms.

stems or leaves some distance from the ground. The larva transforms to a brownish pupa within the cocoon and the flies of the second brood emerge in late June or early July. Sometimes there is a small third brood. The winter is passed in the cocoon, probably in the larval condition.

Treatment.

The larvæ may be readily destroyed by an early application of either 1 pound Paris green or 4 pounds arsenate of lead in 100 gallons of water. Later, when the fruit is near maturity, fresh hellebore should be used at the rate of 4 ounces in 2 or 3 gallons of water or, as a dry application, 1 pound in 5 pounds of flour or air-slaked lime.

The Green Currant Worm

Gymnonychus appendiculatus Hartig

This is the so-called native currant worm, but as it too was doubtless imported from Europe the name should be discarded. It is widely distributed throughout the Northern states and Canada, and while sometimes destructive in the West, it rarely attracts attention in the East.

The adults are smaller than in the preceding species, and in both sexes the body is black. The flies appear in the spring just as the leaves are unfolding, and the female inserts her whitish eggs into the edge of the currant or gooseberry leaf between the two outer layers. The egg swells considerably before hatching and produces a distinct blister-like elevation of the epidermis of the leaf. They hatch in about five days, and the young larvæ feed singly on the edge of the leaf, not in colonies, as in the preceding species. The larvæ are of a uniform light green color and have blackish heads; they lack the black spots characteristic of the foregoing species and are considerably smaller. They become full-grown in about 12 days and spin small brownish cocoons beneath leaves or trash or slightly below the surface of the ground; the flies emerge in about a week. In New York there may be four or five broods, each generation requiring about twenty-five days from egg to adult.

Whenever this currant worm becomes troublesome, it may be controlled by spraying with either 1 pound Paris green or 4 pounds arsenate of lead in 100 gallons of water at the first appearance of the worms. When the fruit is ripening use hellebore, 4 ounces in 2 or 3 gallons of water, or dry 1 pound in 5 pounds of flour or air-slaked lime.

References

Walsh, Pract. Ent. I, pp. 122–124. 1866.
Saunders, Rept. Ent. Soc. Ont. f. 1871, p. 34. 1872.

The Gooseberry Span-worm

Cymatophora ribearia Fitch

Although generally distributed throughout the Atlantic and Middle states and eastern Canada, this measuring-worm becomes troublesome only occasionally. It attacks gooseberry, red and black currant and the blueberry.

The eggs which have remained on the branches through the winter hatch in May about the time the leaves become fully expanded. The larvæ are measuring-worms or loopers, whitish in color with yellow stripes on the back and sides and conspicuously marked with numerous black spots of various sizes. They begin feeding on the tips of the leaves, and, if numerous, soon strip the bush and sometimes attack the tender branches. When disturbed they let themselves down by a silken thread. They become full-grown in three or four weeks and are then an inch or less in length; they leave the bush, enter the ground a short distance and transform to a nearly black pupa about $\frac{3}{16}$ inch in length. The moths emerge in about two weeks, the last of June in New York, and deposit their small, beautifully sculptured, elongate oval, dull yellowish-gray eggs on the bark of the branches more often near the base of the plant. The eggs do not hatch till the following spring, there being but one generation a year. The female moth has an expanse of over an inch, is of a pale yellow color and has the fore wings crossed by two irregular interrupted bands of brownish; the outer band continues across the hind wing. In the male the yellow is darker and the markings more distinct.

Hellebore has not been found effective against this pest, but 4 pounds arsenate of lead, or 1 pound Paris green in 100 gallons of water may be used effectively while the larvæ are small. When nearly full-grown they are not easily poisoned, and recourse must be had to the more tedious and expensive operation of hand-picking.

REFERENCES

Fitch, 3d Rept. State Ent. N. Y., pp. 427–428. 1856.
Saunders, Rept. Ent. Soc. Ont. f., 1874, pp. 18–19. 1875.
Riley, 9th Rept. Ins. Mo., pp. 3–7. 1877.

The Pepper-and-salt Currant Moth

Lycia cognataria Guenée

Currants and gooseberries are sometimes defoliated by a geometrid caterpillar or measuring-worm which, when full-

grown, is nearly two inches in length; it varies in color from sea-green to brownish-gray or brownish-black and is marked with indistinct lines and spots of green and yellowish. When at rest it clasps a twig with the posterior pairs of legs and holds the body extended

Fig. 305. — The pepper-and-salt currant moth ($\times 1\frac{1}{8}$).

rigidly outward; it is then easily mistaken for a dead twig, which it closely resembles in form and color (Fig. 306). It is a general feeder and rarely becomes a pest on currants. This species occurs in the Eastern states and Canada.

The caterpillars become full-grown in early July and transform to a dark brown pupa in the ground. At least part of the pupæ give rise to moths the same season, about the middle of August, but it is quite probable that some of this brood hold over till the following spring. The rather heavy-bodied moths (Fig. 305) have an expanse of about two inches; the wings are gray, dotted with dark brown and crossed by two wavy bands of mahogany brown on the outer third. Moths of the spring brood appear in May. The eggs are about $\frac{1}{30}$ inch in length, cylindrical with rounded ends, and the surface is beautifully marked with rows of hexagonal depressions. A single moth has been known to lay about 500 eggs.

Fig. 306. — Caterpillar of the pepper-and-salt currant moth on a plum branch; it resembles a small dead twig ($\times 1\frac{1}{3}$).

Arsenical poisons as used against the imported currant worm will hold this insect in check.

REFERENCE

Lintner, 2d Report State Ent. N. Y., pp. 97–101. 1885.

THE FOUR-LINED LEAF-BUG

Pœcilocapsus lineatus Fabricius

The four-lined leaf-bug is a native of America and ranges from Canada to Georgia and westward to the Rocky Mountains.

It has been most troublesome in New England, New York and Michigan. It has a wide range of food-plants, but has attracted most attention as an enemy of the currant, gooseberry, mint, parsnip, Weigelia, dahlia and rose.

The insect usually makes its first appearance about the middle of May on the newest, tenderest terminal leaves. The insects are then so small and active in hiding themselves that they are not apt to attract attention. Their work, however, soon becomes apparent. The insect inserts its beak into the leaf and sucks out the green pulp of

FIG. 307. — A gooseberry leaf injured by the four-lined leaf-bug.

the interior within a small area bounded by the little veinlets. As the insects increase in size they suck out the pulp from larger areas. The injured portions of the leaf turn brown and die and give the leaves a characteristic spotted appearance (Fig. 307). The spots often coalesce and the whole leaf turns brown, curls up and dies. The growth of the shoot is checked, and the terminal portion dies. In cases of severe infestation the whole field has the appearance of having been scorched by fire. On currant and gooseberry the insect confines its attacks to the leaves, but on dahlia and rose it attacks the buds.

FIG. 308. — The four-lined leaf-bug (× 2).

The newly hatched nymph is about $\frac{1}{20}$ inch in length, of a bright vermillion red color with large black spots on the thorax. In the course of its development, which requires from 17 to 20 days, the nymph passes through five stages. The full-grown nymph is of a bright orange-yellow color and about $\frac{1}{4}$ inch in length. The black wing-pads a broad yellowish-green stripe near the outer margin and extend about halfway to the tip of the abdomen. At the fifth molt the winged adult appears, about the middle of June in New York, and is about $\frac{1}{3}$ inch in length; the general color of the body is bright orange-yellow with four black stripes on the thorax and wing-covers (Fig. 308); the legs and the areas between the black stripes are dark apple-green, changing to lemon-yellow after death.

FIG. 309. — Currant stem showing white egg-clusters of the four-lined leaf-bug.

The adults are provided with a beak, and feed in the same manner as do the nymphs. They are more voracious, however, and do more damage. They are shy, very active when alarmed and very difficult to capture. Egg-laying begins about a week after the adults appear and is completed early in July. The female is provided with a thin, blade-like obliquely pointed ovipositor by means of which she cuts a slit about $\frac{1}{8}$ inch long in the bark of the soft, tender growth in which she deposits 6 or 8 light yellow eggs. Each egg is $\frac{1}{16}$ inch in length, smooth, slightly curved, and has the upper third capped by a white, finely striated portion (Fig. 310). With the growth of the surrounding tissue the eggs are usually forced out of the stem somewhat, so that

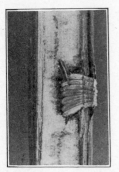

FIG. 310. — Currant stem split to show eggs and the larva of a parasite feeding on them ($\times$ 6).

about one half or more of the white portion projects from
the slit, making it rather easy to locate the eggs (Fig. 309).
There is but one generation a year.

Treatment.

The nymphs may be destroyed while small by a thorough
application of kerosene emulsion diluted with five parts of
water. This is probably the most practicable means of fighting
the pest while in the nymphal stage. The pruning and burning
of the tips in which all the eggs are laid is an efficient means
of keeping the pest in check. When occurring on herbaceous
plants, probably the best method is to capture the bugs by
jarring them into a dish partly filled with kerosene and water.

REFERENCE

Cornell Agr. Exp. Sta. Bull. 58. 1893.

THE CURRANT PLANT-LOUSE

Myzus ribis Linnæus

In the Northern states and Canada the foliage of currants
is very commonly distorted and discolored by the presence of
yellowish-green plant-lice on the under side of the leaves. Red
currants are most subject to attack, but black currants and
gooseberries are sometimes infested. The insect is a native
of Europe, where, in addition to the cultivated varieties, it also
infests the wild alpine currant. It was probably introduced
into America in the early part of the nineteenth century, but
did not attract attention until about 1857.

The shining black, cucumber-shaped eggs, attached to the
bark of the new growth, hatch soon after the leaves open. The
young lice crawl to the leaves and begin feeding on the under
surface ; these lice hatched from the eggs are all females and
are known as stem-mothers. When mature they give birth to
living young, and each individual is soon surrounded by a nu-

merous progeny. Through-
out the summer only fe-
males (Fig. 311) are pro-
duced, and the young are
born alive. At first nearly
all of the lice are wingless,
but as the leaves become
crowded, winged females
develop and migrate to
other bushes.

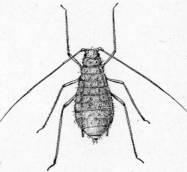

FIG. 311. — The currant plant-louse, wingless viviparous female. Enlarged.

The lice become abun-
dant by the middle of
May in New York, and
often cover the entire under surface of the leaves (Fig. 313).
The irritation of their combined punctures causes the leaf

FIG. 312. — Currant leaves curled by plant-lice.

to become badly curled and distorted so that pocket-like
cavities are formed on the under side (Fig. 312). The upper

surface turns a bright red, variegated with yellow and green. Later the badly injured leaves may fall from the bushes and thus prevent the fruit from maturing properly. The fruit is also injured by a black fungus which grows on the sticky substance, known as honey-dew, secreted by the lice.

FIG. 313. — A colony of currant plant-lice on the underside of a leaf.

After the middle of July the lice become greatly reduced in numbers, owing to the attacks of their numerous predaceous and parasitic enemies, but a few females are able to survive, and at the approach of cold weather give rise to true males and females. The latter deposit the winter eggs on the twigs during the latter part of October.

Treatment.

The currant plant-louse is not an easy insect to control, owing to the way in which it is protected in the pocket-like cavities of the curled leaves. These soft-bodied lice are easily killed by ordinary contact insecticides, such as kerosene emulsion, soap solutions and tobacco extracts; the difficulty is in reaching them. To be effective, the spraying must be done with great thoroughness soon after the hatching of the eggs and before the leaves curl; an upturned nozzle should be used so as to hit the leaves from beneath.

REFERENCE

N. Y. (Geneva) Agr. Exp. Sta. Bull. 139, pp. 660–663. 1897.

THE GOOSEBERRY MIDGE

Dasyneura grossulariœ Fitch

The fruit of the gooseberry is sometimes destroyed by a small, bright yellow maggot which feeds in the pulp and causes the berry to turn red prematurely, decay and fall to the ground. The parent fly is a delicate midge resembling a mosquito in form, about $\frac{1}{10}$ inch in length, with a pale yellow body and legs and with black eyes and blackish antennæ. The female apparently punctures the skin of the fruit with her ovipositor and inserts the egg directly into the pulp.

Pupation takes place within the decayed berry, and the flies emerge in the latter part of July. The further history of this insect is unknown.

The gooseberry midge has never become a serious pest. No better remedy has been suggested than to collect and destroy the infested berries before the flies have had time to mature.

REFERENCE

Fitch, 1st Rept. State Ent. N. Y., p. 176.   1855.

THE GOOSEBERRY FRUIT-WORM

Zophodia grossulariœ Packard

Gooseberries and currants are subject to the attacks of a greenish caterpillar (Fig. 314) with a brownish head $\frac{3}{4}$ inch in length when full-grown, which feeds within the fruit and causes it to color prematurely and either dry up or fall to the ground and decay (Fig. 315). While ordinarily not a serious pest, it has been known to destroy almost the entire crop in certain places, particularly in the West.

The grayish moths have an expanse of nearly an inch; the fore wings are crossed by darker lines, and there is a row of

2 A

small blackish dots near the outer margin. The female is said to deposit her eggs on the fruit, but no one seems to have described the egg. The young larva enters the partly grown

berry and feeds on the pulp, casting out the excrement through the opening in the skin of the fruit by which it entered. It will sometimes enter several berries in succession, and often webs together several berries with a silken thread. When full-grown it descends to the ground and transforms to a

Fig. 314. — The gooseberry fruit-worm. Knight photo.

pupa within a brownish oval cocoon beneath dead leaves or other trash. The winter is passed as a pupa, and the moths emerge the next spring soon after the fruit has set.

The caterpillars are very active, and when alarmed will wriggle out of the berry and hang suspended by a silken thread only to return to the fruit when the danger is passed.

Treatment.

The control of this pest has not been worked out, and

Fig. 315. — Gooseberries injured by the gooseberry fruit-worm. Knight photo.

nothing better than hand-picking of the infested berries has been suggested. Care must be taken in collecting the injured fruit that the caterpillars do not crawl out and escape. While satisfactory in a small garden, hand-picking is too expensive to be practicable under commercial conditions. If poultry are allowed to run in the field after the crop is harvested, they will doubtless devour many of the pupæ in their hibernating quarters beneath trash on the ground.

REFERENCES

Riley, 1st Rept. Ins. Mo., pp. 140–142.   1869.
Saunders, 7th Rept. Ent. Soc. Ontario, for 1876, pp. 39–40.   1877.

THE YELLOW CURRANT FRUIT-FLY

Epochra canadensis Loew

Throughout the Northern states and Canada the currant and gooseberry crop is often seriously injured by a small white maggot which feeds within the fruit. The loss occasioned by this insect seems to be greater in the West, particularly in mountainous regions where there is an abundance of wild berries in which flies may breed undisturbed. Gillette states that in Colorado this fruit-fly is the most serious insect enemy of the currant and gooseberry.

The pale yellowish flies, about as large as the house fly, with green eyes and banded wings, appear in May in the Eastern states, become abundant about the middle of June, and finally disappear after having been on the wing about a month. The female fly inserts her elongate, whitish egg under the skin of the unripe fruit through a puncture made with her sharp extensible ovipositor. The egg is about $\frac{1}{25}$ inch in length, white, elongate oval and provided at one end with a short pedicel. Several eggs may be deposited in a single berry, each in a separate puncture. Each female is capable of laying about 200 eggs.

On hatching, the young maggot may burrow for some distance just beneath the skin before entering the pulp. It then attacks the immature seeds, feeding on the kernel. Infested berries may be distinguished by having a discolored spot either around the puncture or over the place where a maggot has been feeding on the seeds. They color prematurely and usually fall to the ground and decay. In about three weeks the maggots become full-grown; they are then over $\frac{1}{4}$ inch in length, white, with black mouth parts. They then leave the fruit through a ragged hole in the skin, sometimes while it is hanging on the bush, but more often after it has fallen to the ground, and go into hibernating quarters a short distance in the ground or beneath rubbish. The winter is passed in a broadly oval, straw-colored puparium about $\frac{1}{5}$ inch in length. There is only one generation a year.

No practicable method of controlling this insect in large plantings has been suggested. In the garden it might be feasible to collect and destroy the infested berries either before they fall or very soon afterwards. Where poultry are allowed to run under the bushes, they may be able to find and destroy many of the puparia.

REFERENCES

Me. Agr. Exp. Sta. Rept. for 1895, pp. 111–124.
Paine, Psyche, XIX, pp. 139–144. 1912.

THE DARK CURRANT FRUIT-FLY

Rhagoletis ribicola Doane

In Washington and neighboring states currants and gooseberries are also subject to the attacks of a species of fruit-fly closely related to the preceding. The adult is only about half as large as the house fly, black with four yellow stripes on the thorax and a large spot on the scutellum yellow; the head is yellow with greenish eyes, the legs are yellow and the wings are crossed by four brown bands.

The flies are on the wing from the middle of June till the middle of July. The female deposits her eggs just beneath the skin of the berry, and the maggot becomes full-grown in three or four weeks. They enter the ground a short distance, or find protection beneath rubbish and pass the winter in a brownish or black puparium. The habits and life history of this species are very similar to the foregoing, and the means of control are the same.

REFERENCE

Wash. Agr. Exp. Sta. Bull. 36, pp. 3–6. 1898.

THE CURRANT-STEM GIRDLER

Janus integer Norton

In the Northern states and Canada currant bushes are often injured by a sawfly, which, after depositing her eggs in a cane, girdles the tip, causing it to wilt and drop. It is a native insect which probably fed originally on the wild currant, but did not attract attention as an enemy of the cultivated varieties until about 1888.

The sawflies emerge from the middle to the last of May in New York; both sexes have shining black bodies and light brownish-yellow legs (Fig. 316). In the male nearly all of the abdomen is of a brownish-yel-

FIG. 316. — Adult male and female of the currant-stem girdler (× 2½).

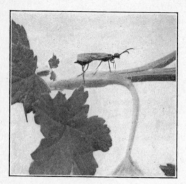

Fig. 317. — The currant-stem girdler ovipositing.

low color, while in the female the front half of the abdomen is reddish-orange, and the rest is black. The female is about $\frac{1}{2}$ inch in length, the male somewhat smaller. The former is provided with a stout, sharp saw-toothed ovipositor, which when exserted extends at a right angle beneath the abdomen (Fig. 317). By means of this ovipositor the female punctures a cane a few inches from the tip and inserts the elongate oval, yellowish-white egg into the pith (Fig. 318). After the egg is deposited she walks up the shoot for from one half inch to an inch and deftly girdles the cane with her ovipositor. Sometimes the girdling is so complete that the tip falls at once, but usually a portion remains uncut and the tip may remain attached for some time, especially if the shoot is a large, vigorous one (Fig. 319). This killing of the tip of the cane seems to be necessary for the development of the egg and grub.

The eggs hatch in about eleven days. The grubs feed almost entirely on the pith, which they tunnel out to a distance of not over six inches, leaving the burrow packed full of excrement behind them. The borer becomes full-grown about the first of September and cleans out the lower end of its burrow for the distance of about three fourths inch and then eats a passageway out to the outer bark, which soon dies and shrinks over

Fig. 318. — Egg of the currant-stem girdler in position. Enlarged.

this point. It then surrounds itself with a silken cocoon within which it remains as a grub all winter. The change to a pupa takes place in the spring, and the adult insect emerges a few days later.

The currant-stem girdler cannot be reached at any time or in any way with a spray. Fortunately, however, its habits are such that it can be easily controlled by other means. The girdling habit of the adult insect which causes the young shoot to wilt, die, and drop off in May makes it easy to determine whether the pest is present or not. Since the egg is embedded in the shoot less than an inch below where the girdling is done, and as the grubs rarely tunnel down more than six inches, if the injured shoots are cut off at least eight inches below the

Fig. 319. — Girdled portion of a stem, much enlarged, to show the character of the girdle.

girdle and burned, the insect will be effectively controlled. If the work is performed in May or June soon after the girdling is done, only two or three inches of the tips need be cut off. The cutting and burning of about eight inches of the tips of the injured shoots at any time of the year, even in winter, will prove an effective remedy for this pest.

REFERENCES

Marlatt, Ins. Life, VII, pp. 387–390. 1895.
Cornell Agr. Exp. Sta. Bull. 126, pp. 41–53. 1897.

The Walnut Scale

Aspidiotus juglans-regiæ Comstock

Mature female scales of this species are pale grayish-brown in color with the reddish-orange exuvial spot one side of the center, and they are larger than the San José scale, being about $\frac{1}{8}$ of an inch in diameter. The species is oviparous, hibernated adult females laying eggs early in the spring, and eggs for another brood are laid in June. There are two and possibly three generations of this scale insect in the South, yet it rarely occurs in injurious numbers. We have seen currant stems incrusted with it, and it also occurs on apple, pear, cherry, peach, apricot and plum, besides on its only food-plant in California, the English walnut. It is widely distributed over the United States and occurs in Canada.

This walnut scale will doubtless succumb to thorough applications of the winter washes recommended for the San José scale.

Other Currant and Gooseberry Insects

CHAPTER XII

STRAWBERRY INSECTS

STRAWBERRY plants are at the most short-lived; they are low growing and are usually cultivated in closely set rows. Strawberry growing is more akin to the raising of field crops than to the cultivation of other fruits. Likewise in the control of strawberry insects less reliance is placed on spraying and more attention is given to crop rotation, fall plowing, clean cultivation, and similar practices. The one-crop system of strawberry culture as now practiced by the majority of commercial growers greatly simplifies the problem of insect control. This is especially true in the case of white grubs, root worms and other under-ground insects.

THE STRAWBERRY LEAF-ROLLER

Ancylis comptana Frölich

Throughout the Northern states and Canada, from Colorado eastward, strawberries are often seriously injured by a small greenish or brownish caterpillar which folds the two halves of the leaflets together and feeding within the shelter so formed causes them to turn brown and die. In years of great abundance the injury may be very severe; the foliage is destroyed and the fruit fails to mature.

The parent moth measures about $\frac{3}{5}$ inch across the expanded wings; its general color is light reddish-brown and the fore wings are marked with wavy bands of white and darker brown.

The moths appear in the strawberry fields in early May in
New Jersey and in April in southern Missouri. The female
deposits her pale green, round or slightly oval, strongly flattened
eggs singly on the underside of the leaves. One observer says
they are laid on the upper side along the mid-rib. The eggs
hatch in about a week and the young caterpillars crawl to the
upper surface, where they feed for a day or two openly on the
upper epidermis without any protective covering. Within
a few days the larva begins to draw the two halves of the leaflet
together with silken threads and finally produces a complete
fold, within which it finishes its growth in about a month.
When full-grown it is about $\frac{1}{2}$ inch in length, varies in color
from yellowish to greenish-brown and has the head and cervical
shield shining brown. It transforms to a pale brownish pupa,
$\frac{5}{16}$ inch in length, within the folded leaf, and in about ten days
the moth emerges. From 42 to 50 days are required for the
development from egg to moth. In New Jersey there are three
generations a year, but the later broods greatly overlap. Farther
north there are only two generations a year, while there is some
evidence of a fourth brood in Kentucky. The insect hibernates
both as a larva and as a pupa. At the approach of cold weather
some of the partly grown caterpillars desert the leaves and seek
shelter beneath trash or the mulch, returning to the leaves to
complete their growth the following spring; while those that
are mature transform to pupæ and remain in that condition
in the folded leaves until the following spring, when they give
rise to the first brood of moths.

The strawberry leaf-roller also attacks the blackberry and
raspberry, being especially abundant on these plants during the
latter part of the season. On strawberries it is usually the first
brood that causes the greatest loss. The caterpillars of the
later broods are, as a rule, less numerous and, owing to the greater
quantity of foliage on which to feed, cause less apparent injury
to the plants.

Means of control.

Experiments in New Jersey have shown that the strawberry leaf-roller can be effectively controlled by a single, timely application of arsenate of lead, 5 pounds in 100 gallons of water. The poison should be applied within a week after the first appearance of the moths and just before the young larvæ begin to fold the leaves. Spraying after the leaves are folded will do little or no good.

In some parts of the country the leaf-roller is controlled by burning over the strawberry field soon after the crop is harvested.

Fig. 320.—Larva of the obsolete-banded strawberry leaf-roller beginning to roll a leaf.

Fig. 321.—Full-grown larva of the obsolete-banded strawberry leaf-roller (× 3).

In this way practically all the larvæ and pupæ in the folded leaves are destroyed.

Strawberry beds that are to be abandoned should be plowed under directly after the picking of the last crop and not allowed to remain as breeding places for the moths.

REFERENCES

Forbes, 13th Rept. State Ent. Ill., pp. 87–93. 1884.
Ky. Agr. Exp. Sta. Bull. 31, pp. 13–16. 1890.
N. J. Agr. Exp. Sta. Bull. 225, pp. 17–23. 1909.

The Obsolete-banded Strawberry Leaf-roller

Archips obsoletana Walker

Occasionally strawberry beds are seriously injured by a small olive-green caterpillar which folds the leaves or webs them together so as to destroy the foliage and prevent the ripening of the fruit. Outbreaks have been reported in Illinois and New York only, but the insect is widely

Fig. 322.— Pupa of the obsolete-banded strawberry leaf-roller (× 5).

Fig. 323. — Male (above) and female (below) moths of the obsolete-banded strawberry leaf-roller.

distributed throughout the eastern United States from Texas to Massachusetts.

Where or in what stage the insect passes the winter is not known. In May the young caterpillars appear on the leaves, where they feed at first on the underside of the leaves, skeletonizing small areas; they first live in a little tube formed by tying the leaf hairs together with silk near a vein. In a few days the caterpillar crawls to the upper surface and begins to fold the

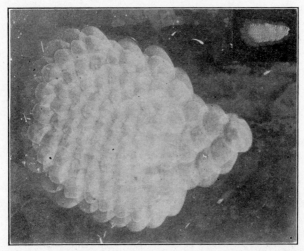

FIG. 324. — Egg-cluster of the obsolete-banded strawberry leaf-roller, much enlarged, with another cluster natural size in upper right-hand corner.

leaflet by spinning silken threads above its body from one side to the other across the mid-rib and thus gradually drawing the edges of the leaflet together (Fig. 320). Within this protecting roll the caterpillar lives, feeding upon the leaf and often joining other leaves to it. Sometimes the caterpillar forms a similar protection by drawing together the blossoms and forming fruits, which it eats (Fig. 325).

In New York the first brood of caterpillars (Fig. 321) mature about the first

FIG. 325. — Young fruits tied together and ruined by the caterpillars.

week of June and transform to slender dark brown pupæ, about ½ inch in length, within the folded leaf (Fig. 322). In about ten days the moths emerge (Fig. 323). The general color of the moth varies from wood-brown to russet; the fore wings are crossed obliquely by a broad, dark brown band and have a large spot of the same color near the tip. The thin, oval, light lemon-yellow eggs are laid in clusters of more than a hundred, overlapping each other like shingles on

a roof (Fig. 324). Just where the moth places her eggs in the field has not been determined. The eggs hatch in about ten days. In New York there are three broods a year, caterpillars developing in May, July and December.

Fig. 326. — How the caterpillars destroy the leaves which they roll.

Means of control.

The measures suggested for use against the strawberry leaf-roller on page 363 would doubtless be effective against this species.

REFERENCE

Cornell Agr. Exp. Sta. Bull. 190, pp. 145–149. 1901.

THE BLACK-MARKED STRAWBERRY SLUG

Empria maculata Norton

In the Northern states east of the Rocky Mountains and in Canada the foliage of the strawberry is occasionally attacked by the greenish larvæ of a small dark colored sawfly. Outbreaks have been reported from Missouri, Illinois, Indiana and Canada.

The adult is a black-bodied, four-winged fly about $\frac{1}{5}$ inch in length and has a row of whitish spots on each side of the abdomen. The flies emerge from the ground in early spring and the female inserts her pure white, elongate, compressed eggs in the petioles of the leaves. The eggs hatch in about two weeks and the yellowish or greenish larvæ attract attention in May by eating small round holes in the leaves. When abundant they may completely destroy the foliage in a few days. Much of the feeding is done at night; during the day the larvæ remain curled up on the underside of the leaves or hide at the base of the plant. They become full-grown by the last of June in Illinois and are then about $\frac{3}{4}$ inch in length, pale greenish to grayish-yellow in color; the head is yellow, marked with at least three large dark spots. When mature the larva enters the ground and constructs an oval cocoon composed of earth cemented together with a gummy substance within which it passes the winter as a shortened, thickened larva, the so-called pre-pupa. Pupation takes place in early spring and the adults emerge soon after.

As a rule, there is only one generation a year, but there is evidence to show that occasionally a partial second brood may develop in the more southern part of the insect's range.

Means of control.

The sawfly larvæ are readily killed with arsenicals, and if the application is made before the fruit is more than one half grown, they may be used with perfect safety. Paris green, 1 pound in 150 gallons of water, with the addition of three pounds of fresh lime, has given good results, but as there has been some complaint of burning of the foliage with this poison, arsenate of lead, 4 pounds in 100 gallons of water, will probably be found more satisfactory. It should be used at the first appearance of the larvæ, before the berries are more than one half grown. After the fruit has begun to ripen use hellebore, 4 ounces in 2 gallons of water.

REFERENCES

Riley, 9th Rept. Ins. Mo., pp. 27–29.   1877.
Forbes, 13th Rept. State Ent. Ill., pp. 71–76.   1884.
Forbes, 14th Rept. State Ent. Ill., pp. 77–78.   1885.
Iowa Agr. Exp. Sta. Bull. 18, pp. 512–514.   1892.
Ohio Agr. Exp. Sta. Bull. 68, pp. 33–35.   1896.
Mo. Agr. Exp. Sta. Bull. 54, pp. 192–202.   1901.

The Green Strawberry Slug

Empria ignota Norton

This species is closely related to the one last treated, but differs from it somewhat in the details of its life history.   The injury inflicted is, however, practically the same.   It has been reported as troublesome in Iowa, Illinois and Indiana.

The flies appear in the spring somewhat earlier than the preceding species and deposit their eggs singly just beneath the lower epidermis of the leaf, thereby causing small light-colored blisters.   Three or four eggs are laid in a leaf.   As soon as hatched the young larvæ begin to eat small holes in the leaves, and if abundant, may strip the plants of their foliage, leaving the fruit stunted and unable to ripen properly.   They molt four times, becoming full-grown in May or early June.   They are then over one half inch in length, deep green in color with obscure blackish dorsal and lateral stripes; the head is uniform yellowish-brown without the distinctive blackish spots of *E. maculata* described above.   When full-grown the larva enters the ground an inch or so and forms a frail earthen cocoon lined with silk, within which the insect remains in a shortened and thickened condition till the following spring, when the transformation to the pupa and adult takes place.

Treatment.

The green strawberry slug may be controlled by the measures suggested for the species last treated.

REFERENCES

Mally, Insect Life, II, pp. 137–140. 1889.
Mally, Insect Life, III, pp. 9–12. 1890.

In Colorado a sawfly (*Emphytus gillettei* MacGillivray) with habits almost identical with those of the above has been reported as locally destructive. The larva is nearly an inch in length when full-grown, green above and cream colored beneath; the head is yellowish with a brown patch above.

REFERENCE

Johnson, Rept. Ent. Col. 1903, pp. 13–14.

The Strawberry Whitefly

Aleyrodes packardi Morrill

This near relative of the common whitefly of the greenhouse is often found on strawberries growing out of doors, but has been reported as noticeably destructive only in one instance, in southeastern New York.

The insect passes the winter as a minute, smooth, oval, metallic bronze-colored egg attached to the underside of the leaf by a short stalk. The winter eggs hatch in early spring into minute, flat, pale green, lice-like insects which are able to crawl some distance over the surface of the leaf. The insect soon settles down, punctures the leaf with its bristle-like mouth parts through which it feeds on the sap. It remains stationary, soon loses the use of its legs and secretes a covering of wax which gives it the appearance of a scale insect. It molts three times and then assumes the so-called pupa form. In this stage the insect appears as a delicate, greenish-yellow, oval wax box with perpendicular sides, about $\frac{1}{30}$ inch in length, and has the dorsal surface ornamented by long curved waxen rods. In about

2 B

a month from the hatching of the egg the minute, mealy white, four-winged flies appear to lay eggs for another brood.

The winter eggs hatch in early spring and those nymphs that escape the spring frosts give rise to adults in May; breeding continues until cold weather. As a rule, the insects are most abundant the latter part of the season. The nymphs secrete a sweet sticky substance known as honey-dew, which collecting on the leaves serves as a medium for the growth of a black fungus, which gives infested plants a characteristic sooty appearance.

Methods of control.

Fortunately the strawberry whitefly is rarely abundant enough to injure the crop to any great extent. The nymphs are easily killed by the ordinary contact insecticides, soap solution, nicotine, etc., but as they are found exclusively on the underside of the leaves it would be a difficult matter to hit them. It might be possible, however, to do effective work by using an upturned nozzle attached to the end of a short rod to be used as a handle.

REFERENCES

Cornell Agr. Exp. Sta. Bull. 190, pp. 155–158.   1901.
Morrill, Can. Ent. XXXV, pp. 25–35.   1903.
Mass. Agr. Exp. Sta. Tech. Bull. 1, pp. 53–62.   1903.

THE STRAWBERRY FLEA-BEETLE

Haltica ignita Illiger

This small, green, coppery or blue flea-beetle about $\frac{1}{6}$ inch in length is widely distributed throughout the country from Hudson Bay to Florida and Texas. The beetles emerge from hibernation in early spring and feed voraciously on the tender foliage of a number of wild and cultivated plants, including the grape and strawberry. They have been found most destructive to strawberries in the South, Florida and Texas. The injury to strawberry plants is sometimes very severe; the beetles appear in immense numbers and completely riddle the leaves.

The beetles deposit their eggs on leaves of the evening primrose and related plants. The larvæ feed on the leaves and seedpods, going into the ground to transform. There is only one brood a year in the North and two or three in the South.

Treatment.

The beetles may be driven away by thorough applications of Bordeaux mixture. By adding arsenate of lead, 4 to 8 pounds in 100 gallons, it is possible to kill a few of the beetles, but probably not enough to pay for the trouble. As the only known food-plants of the larvæ are the evening primrose and its allies, such weeds should not be tolerated in the vicinity of strawberry beds.

REFERENCE

Chittenden, U. S. Bur. Ent. Bull. 23, pp. 70–78. 1900.

The Imbricated Snout-beetle

Epicærus imbricatus Say

Strawberry plants are sometimes defoliated by this greenish-brown snout-beetle $\frac{3}{8}$ to $\frac{1}{2}$ inch in length. The wing-covers are crossed by two irregular light bands, more distinct in the males. The greater part of the insect is clothed with small appressed scales which are imbricated or overlap, like the scales of a fish, hence the common name. The insect is found east of the Rocky Mountains, except in the extreme North and South. It is a general feeder in the adult stage, attacking, among others, apple, peach, pear, cherry, raspberry, gooseberry, onions, beets and cabbage, in addition to a large variety of wild plants.

In confinement the female deposits her smooth elongate, dull yellow eggs in clusters on a leaf and then glues another leaf or a portion of the same leaf over them. The larva is a root feeder, but its natural food-plant is unknown.

Treatment.

It is probable that the beetles could be either killed or driven away from strawberry plants by a thorough application of arsenate of lead, 5 to 8 pounds in 100 gallons of water. Of course, it would not be safe to use this poison after the fruit had attained much size.

REFERENCE

Chittenden, U. S. Bur. Ent. Bull. 19, pp. 62–67. 1899.

THE STRAWBERRY WEEVIL

Anthonomus signatus Say

The size of the strawberry crop in the country east of the Rocky Mountains is often greatly lessened by the attacks of a small reddish-brown to black weevil which after laying an egg in the flower bud causes it to fall by cutting the pedicel. In badly infested localities losses of 50 to 60 per cent of the crop are not uncommon. Fortunately its attacks are of an intermittent nature; after two or three years of abundance in a locality the weevil usually disappears and does not again attract attention for a much longer period.

The insect hibernates in the beetle stage, under rubbish, particularly in wood lots or hedge rows adjoining strawberry fields. The beetles (Fig. 327) are only about $\frac{1}{10}$ inch in length and vary from almost black to reddish brown, with the head and thorax more or less black and with a large black spot on each wing-cover.

The adults forsake their winter quarters in the spring and appear in the strawberry fields as soon as the blossom buds put forth. After feeding to a slight extent on immature pollen procured by puncturing the blossom buds the female deposits her eggs singly in the interior of the nearly mature but unopened buds of the staminate varieties. She first punctures the floral envelope of the bud with her snout and then turning around inserts the smooth, oval, pale yellow egg, about $\frac{1}{50}$ inch in diameter, through this puncture into the interior of the bud, where it lies upon the unopened stamens. She then crawls down the stem of the bud and girdles it so that the bud either falls to the ground at once or hangs a few days attached by a few shreds of tissue. This operation serves to prevent the opening of the bud and thus provides protection to the future grub. Most of the buds fall to the ground within a few days, where they are more liable to remain moist, a condition necessary for the development of the larvæ.

The egg hatches in about a week and the young grub at first feeds almost entirely on the pollen, but later may attack other parts of the interior of the bud. When full-grown the grub is about $\frac{1}{10}$ inch in length, strongly curved and of a white

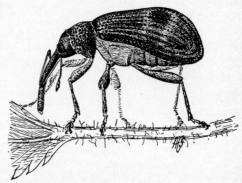

Fig. 327. — The strawberry weevil ($\times$ 15).

or yellowish color. The larvæ become mature in three or four weeks and construct a cell in the frass with which the bud is filled. The pupal stage lasts about a week; thus completing the whole life-cycle in four or five weeks. The new crop of beetles feed for a short time on the pollen of flowers, especially those of the wild bergamot, and then disappear, going into hibernation in midsummer. There is only one generation a year.

The strawberry weevil originally bred in the buds of red-bud or Judas-tree, the wild blackberry, dewberry and strawberry as well as in those of the yellow flowered cinquefoil, but the cultivated strawberry is now the favorite food-plant. The larva requires a diet of pollen and the female as a rule instinctively selects for oviposition the buds of those varieties only which are well supplied with this substance. Varieties with imperfect flowers, that is, lacking stamens, are attacked only to a slight extent.

Treatment.

Practically all of the injury caused by the strawberry weevil is due to the cutting off of the blossom buds by the female in oviposition. Fortunately, as a rule, she confines herself instinctively to the staminate varieties and leaves the plants with imperfect flowers untouched. In view of these facts it has been common to recommend the setting of varieties with imperfect flowers for the main crop and planting only every fifth row to some perfect flowered form in order to insure proper fertilization. This method is said to have been used with success in Maryland and Virginia. Owing to the difficulty of finding imperfect varieties having all the other characters necessary to make them commercially profitable, many growers in infested regions rely on profusely blooming perfect flowered varieties. The weevils attack these, it is true, but owing to the abundance of blossoms, generally enough are left to give a good crop. The greater part of the injury is done within two weeks after the first buds mature on the early varieties. Wherever it is possible to have the main crop come a little later much of the injury can be avoided. In fact, it might be worth while to use very early profusely blooming perfect varieties as a trap crop. A row or two of these plants should be set on the exposed side of the field adjoining wood land or hedge rows in which the beetles hibernate. The beetles will congregate on these rows and deposit eggs there, after which the plants should be mowed, allowed to dry and then burned, or they may be plowed under deeply, thus killing the grubs in the buds. As a rule, it is not advisable to plant strawberries next to wood lots or waste land, or to tolerate hedge rows or overgrown fences in the vicinity. Such cover furnishes excellent winter protection to the beetles and makes their control unnecessarily difficult. Whenever the plants are sprayed with Bordeaux mixture for the control of fungous diseases while they are in bud, arsenate of lead at the rate of 4 pounds to 100 gallons should be added and would probably help to decrease the numbers of beetles. Applied at that time there could be no danger of poisoning the fruit.

Some growers mow their vines soon after picking and, after the tops have dried, scatter straw lightly over the field and then burn it over. In this way many of the grubs are destroyed, but there is considerable danger of injuring the plants unless the work is done very carefully.

Probably the best results in the control of the strawberry weevil can be obtained by clean cultivation, by the destruction of the hibernating quarters of the beetles, and by the planting of varieties with imperfect flowers for the main crop interspersed with a few rows of staminate-flowered plants or by selecting profusely blossoming varieties. The other measures may be used when local conditions make them desirable.

References

Chittenden, Ins. Life, V, pp. 167–186. 1893.
Chittenden, Ins. Life, VII, pp. 14–23. 1894.
N. C. Dept. Agr. Ent. Circ. 12. 1904.
Chittenden, Bur. Ent. Circ. 21, rev. ed. 1908.
N. J. Agr. Exp. Sta. Bull. 225, pp. 8–17. 1909.

THE TARNISHED PLANT-BUG

Lygus pratensis Linnæus

This inconspicuous brownish sucking plant-bug is widely distributed throughout the northern hemisphere, occurring in North America, Europe and Asia. It is a general feeder, attacking a great variety of cultivated and wild plants. In feeding it punctures the buds and tender growing tips with the sharp bristles of its beak, and sucks out the plant juices; at the same time it apparently injects into the wound some

substance poisonous to the plant which kills the surrounding tissue. Peach nursery stock is particularly liable to injury. The bugs attack and kill the tender tips (Fig. 328), causing the tree to throw out lateral branches which are in turn

Fig. 328. — Tip of peach nursery tree injured by the tarnished plant-bug.

similarly injured, producing an overly branched, scrubby tree which cannot be sold as first-class stock. Pear and apple stock are often attacked, but seem able to outgrow the injury more easily than peach.

As a strawberry pest the tarnished plant-bug often causes considerable loss by puncturing the young fruits before the receptacle expands. Berries thus injured remain small and hard and turn dark colored; when the injury is only partial, they become deformed at one side or knobbed at the tip; in either case they are unfit for the market. This injury is known to the growers as buttoning.

Fig. 329. — The tarnished plant-bug ($\times 4\frac{3}{8}$).

The tarnished plant-bug has been known to injure the pear crop severely by puncturing and killing the opening buds and blossoms. It has been known to deposit its eggs in young apples, causing scars which persist as dimples in the mature fruit. Among florists it is recognized as a serious enemy of asters, chrysanthemums and dahlias.

The adult tarnished plant-bug (Fig. 329) is about $\frac{1}{5}$ inch in length, inconspicuously colored, dull yellowish or greenish mottled with reddish-brown. The males are generally much darker than the females. In the North the insect hibernates in the adult state under trash, in stone piles, and along fences; farther south the older nymphs are said to survive the winter. About the time the buds burst, the adults appear on tender foliage of all sorts, where they feed on the sap. They are very active, taking flight at the slightest alarm. The female inserts her eggs (Fig. 330) the full length singly or in small groups in punctures in the tender growing tips or in the petioles and veins of the leaves. In late summer eggs are laid in the flower heads of aster and other composite plants. The egg is a little less than $\frac{1}{25}$ inch in length, flask-shaped, truncate at the

Fig. 330. — Eggs of the tarnished plant-bug in position in a tender peach tip (× 11).

outer end, which is provided with a cap. The time required for the hatching of the egg is about ten days.

The yellowish-green nymphs pass through five stages before attaining wings: the first and second stages are unspotted; the third to fifth are marked with distinct black spots on the thorax and abdomen. The time spent in the nymph state is about one month. There are probably several generations each year, as adults and nymphs of all stages are found from May till heavy frost in the fall.

The tarnished plant-bug has been found a very difficult insect to control, owing to its wide range of food-plants, which includes such common weeds as golden rod, wild carrot, wild asters and mullein, and to the fact that a large part of the injury is done by the adults which are so shy and active that it is

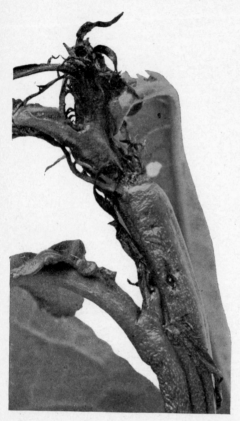

Fig. 331. — Egg of the tarnished plant-bug inserted near the tip of a peach nursery tree; the terminal bud has been killed by the feeding punctures of the bugs. Enlarged.

difficult to hit them with a spray. Much may be done, however, to lessen their numbers by keeping down all weeds, not only in the field itself, but along fences and in other waste land. Stone piles and fences, rubbish heaps, sodded driveways and nearby woodlands furnish hibernating quarters for the adults, and should be avoided whenever possible. The young may be killed by spraying with kerosene emulsion or tobacco extract and soap, but this treatment is not effective against the adults. It has been suggested that in nurseries and strawberry beds the adults may be captured by means of a butterfly net. While this method may be useful in the small garden, it is not adapted for use on a large scale. The control of the tarnished plant-bug in the nursery, on asters when grown for seed and in commercial strawberry fields is still an unsolved problem.

References

Forbes, 13th Rept. State Ent. Ill., pp. 115–135.   **1884.**
Mo. Agr. Exp. Sta. Bull. 47.   1899.
Back and Price, Jour. Ec. Ent. V, pp. 329–334.   1912.

In Florida a small dark brown sucking bug with light-colored wings and brownish legs, *Pamera vincta* Say, attacks strawberries, sucks out the contents of the ovaries and causes a buttoning of the fruit like that produced by the tarnished plant-bug.   A similar injury is caused by the leaf-footed plant-bug (*Leptoglossus phyllopus* Linnæus).   Satisfactory methods of controlling these insects have not been worked out, but the suggestions given above for fighting the tarnished plant-bug may be found of value.

Reference

Fla. Agr. Exp. Sta. Bull. 42, pp. 564–577, 581–583.   1897.

The Strawberry Thrips

Euthrips tritici Fitch

This is the commonest and most widely distributed species of thrips in this country and occurs abundantly in the flowers of almost any wild or cultivated plant.   Destructive outbreaks in strawberry fields have been reported from Illinois and Florida.

The adult thrips is a slender insect about $\frac{1}{20}$ inch in length, usually brownish-yellow in color, although very variable in this respect, and is provided with two pairs of narrow wings margined with a row of long hairs; the immature stages closely resemble the adult in general form, but are wingless and are of a lighter or clearer yellow color.

The thrips appear on the strawberry plants in early spring, and as soon as the buds open are found in the flowers.   Their mouth parts are intermediate between the sucking and biting types.

In feeding they rasp or chafe away the epidermis of the tenderest parts of the flower, and then suck up the sap. The delicate pistils suffer most severely from their attacks; they turn black, wilt, and the development of the ovary is prevented. In severe cases the whole blossom wilts and dries up within a few days after opening; when only part of the ovaries are destroyed, the fruit may mature, but undersized, misshapen and distorted berries only are produced.

The female inserts her minute, whitish, oblong, curved eggs singly in the lower part of the calyx and in the flower stalk. They hatch in about three days. The nymphs pass through three immature stages and acquire wings at the third molt. Only about twelve days are required for the complete life-cycle from egg to adult. There are several generations each year.

Control.

Injury by thrips is usually most severe in seasons of drought, as driving rains destroy great numbers of the insects in all stages of development, except the eggs. Experiments in Florida have shown that this pest can be readily destroyed by thorough spraying with tobacco extract, at intervals as the abundance of the insects require. "Black Leaf 40" tobacco extract should be diluted at the rate of 1 part in 1000 parts of water for this purpose. To make the liquid spread and stick better, 2 pounds of soap should be added to each 50 gallons.

REFERENCES

Fla. Agr. Exp. Sta. Bull. 42, pp. 552–564. 1897.
Fla. Agr. Exp. Sta. Bull. 46, pp. 80–103. 1898.

GROUND-BEETLES

Harpalus caliginosus Fabricius and *H. pennsylvanicus* Dejean

These two common species of the familiar black ground-beetles (Fig. 332) have been reported as injuring the fruit of the strawberry in Pennsylvania, Ohio and Iowa. The favorite haunts of these beetles are under stones and rubbish on the ground, hence the usual mulch on a strawberry bed forms an ideal lurking place for them. The larvæ are said to be predaceous, and the beetles themselves have been generally considered as beneficial insects. While their favorite food consists of other insects, they are often seen in the fall on ragweed, feeding on

Fig. 332. — Male and female of a ground-beetle, *Harpalus caliginosus* (× 2).

the seeds. Both species are attracted to electric lights, where they often occur in immense numbers. Very little is known in regard to the life history or habits of the early stages of these beetles, and no explanation has been offered to account for their sudden appearance in such numbers in strawberry beds.

As strawberry pests these beetles have been known to destroy a crop almost completely in a day or two. They hide during the day by thousands beneath the straw mulch and emerge at dusk to feed on the seeds of the berries. At first only the seeds are eaten (Fig. 333), but later most of the pulp of the ripe berries is devoured; even the green berries are attacked. When only a small part of the pulp is eaten, the berry is ruined, for rot soon ensues. In one case the beetles destroyed $\frac{9}{10}$ of a crop in two days.

Treatment.

Until more is known in regard to the life history and habits of these ground-beetles, it is difficult to suggest methods of control. In England, where a closely related species attacks strawberries, the growers endeavor to protect the vines by sinking dishes in the ground baited with pieces of meat.

Fig. 333. — Strawberries showing the destructive work of ground-beetles; at the extreme right is an uninjured fruit.

REFERENCES

Webster, Can. Ent. XXXII, pp. 265–271. 1900.

Cornell Agr. Exp. Sta. Bull. 190, pp. 150–154. 1901.

THE STRAWBERRY ROOT-LOUSE

Aphis forbesi Weed

This destructive enemy of the strawberry is closely related to and sometimes confused with the well-known melon aphis. It is widely distributed throughout the states east of the Rocky Mountains from Louisiana to Minnesota and New Hampshire, but the most destructive outbreaks have been reported from

Illinois, Ohio, Maryland and Delaware. Its presence in a field is sometimes first indicated by the drying out of plants in certain spots; in other cases the plants generally have an unthrifty look, and the fruit remains small and fails to ripen. It has been found more troublesome on light sandy soils.

The insect passes the winter as shining black elongate oval eggs attached to the leaves and leaf stems of the strawberry plant. These winter eggs hatch in early spring, giving rise to wingless females which, when mature, give birth to living young. These are all females, and reproduction continues agamically throughout the growing season, males not appearing till the advent of cold weather in the fall. At first the lice feed exclusively on the leaves and tender parts of the plants above ground, but about the last of April in Delaware the little brown ant, also known as the corn-field ant (*Lasius niger americanus*) becomes abundant and carries many of the young aphids down to the roots, where colonies are established that soon sap the vitality of the plant. The ants feed on the honey-dew secreted by the aphids and care for them somewhat as we do for our domestic animals. In case the plant dies or the roots become overcrowded, they transfer their "cows" to green pastures on some near-by plant.

Most of the aphids are wingless, but when the supply of food becomes insufficient, winged forms are produced that fly to neighboring fields where they are found by ants, and new colonies are established on the roots. In Delaware these winged females are most abundant in May and in June. The aphids multiply with marvelous rapidity, one generation following another about every two weeks. At the approach of cold weather males and egg-laying females are produced; the latter deposit the winter eggs on the leaves and leaf stems above ground.

Means of control.

Much of the loss caused by the strawberry root-louse can be avoided by setting clean plants on uninfested land. It is not advisable to plant strawberries continuously on the same land; some other crop should intervene. As soon as beds are abandoned, they should be plowed up and not left as centers of infestation for surrounding fields. Burning over the field in early spring before any of the eggs have hatched has been recommended as a satisfactory means of control, and doubtless would be one of value under some conditions.

Whenever possible, plants for setting should be taken from uninfested fields, but when that is not practicable, they should be freed of the aphids before planting. Wait till all the eggs have hatched in the spring, and then dip the plants, roots and tops, in tobacco decoction, or nicotine extract, one part in 1000 parts of water, or they may be fumigated with hydrocyanic acid gas, using 1 ounce per 100 cubic feet of space for 10 minutes.

REFERENCE

Del. Agr. Exp. Sta. Bull. 49, pp. 3–13.   1900.

THE STRAWBERRY CROWN-MOTH

Sesia rutilans Henry Edwards

On the Pacific slope the strawberry, blackberry and raspberry are subject to injury from the attacks of a borer, the larva of a clear-wing moth. The dirty white, brown-headed caterpillar, three fourths inch in length when full-grown, burrows in the crown and causes the plant to wilt, dry up and die.

In the southern part of their range the moths are on the wing during May and June, being most abundant the last of May; in Washington they are abroad in July. The female moth has an expanse of nearly an inch; the body is black, marked with yellow lines and bands; the fore wings are blackish-brown with yellow rays along the veins. The hind wings are transparent with a narrow border of brown-black. The male is smaller and has the transparent areas of the wings larger. The moths are active in the bright sunlight, and when flying are easily mistaken for wasps. The female is said to deposit her egg on the strawberry crown at the base of the leaves. The caterpillar burrows into the crown, at first feeding near the surface, but later eats out the whole interior of the main root, thus killing the plant. It hibernates in the nearly full-grown condition, completes its growth the following spring and pupates in the upper part of its burrow in a slight silken cocoon into which are incorporated bits of frass. When ready to transform to the adult, the pupa works itself part way out of the cocoon, and on the escape of the moth the empty pupal skin is left projecting from the opening of the burrow. There is only one brood a year.

Remedial measures.

Throughout its development the larva, feeding on the interior of the crown, is out of reach of poisons. No better remedy has been suggested than to pull up and burn infested plants before the moths emerge. Strawberry beds which are to be abandoned should be plowed under after harvesting the last crop, and not left as breeding places for the moths. Observations in California indicate that a large percentage of the borers can be killed by submerging the beds for four or five days in winter or early spring. This method may be found of value in irrigated districts.

REFERENCES

Wash. Agr. Exp. Sta. Bull. 35, pp. 13–17. 1898.
Chittenden, Bur. Ent. Bull. 23, pp. 85–90. 1900.

Strawberry crown-miner.

A small lepidopterous caterpillar has been occasionally reported in Illinois and Canada as injuring the strawberry by mining the crown. This insect has been considered as identical with the peach-twig borer (*Anarsia lineatella* Zeller), but is probably a distinct species. The life history of this insect has not been fully worked out. The reddish-pink caterpillar about

2 c

½ inch in length apparently passes the winter in a partly grown condition, inside a silky case in its burrow in the crown; completes its growth the following spring and pupates in May or June. There is apparently only one generation annually.

The Strawberry Crown-girdler

Otiorhynchus ovatus Linnæus

While the larvæ of this beetle often injure strawberry plants by eating off the roots near the crown, whence the name girdler,

it has attracted more attention from the annoying habit the beetles have of invading dwellings in search of shelter. The insect occurs in the Northern states and Canada and is widely distributed in Europe and northern Asia.

Fig. 334. — Beetle of the strawberry crown-girdler (× 9).

The full-grown grub is ¼ inch or less in length, whitish in color with a yellowish head, and is strongly curved. The grubs feed on the roots of the strawberry, cutting them off near the crown, but as a rule do not burrow into the crown like the crown-borer. They also attack the roots of various grasses, white clover, wild strawberry and related plants. Badly infested strawberry plants are killed outright. The time required for the grubs to reach maturity has not been determined. When full-grown the grub constructs a small earthen cell within which it transforms to a whitish pupa and later to an adult. The latter is a very dark brown, almost black, snout-beetle, about $\frac{3}{16}$ inch in length (Fig. 334). The beetles are unable to fly; the wings are absent and the wing-covers have grown together so they cannot be opened for flight. They often congregate in great numbers around the base of the plant, and frequently burrow into the surrounding soil. The beetles feed on the foliage of various plants, and when abundant may seriously injure newly set strawberry plants. The female deposits her minute whitish oval eggs either on the surface of the ground or in her burrows in the soil among the roots. The eggs hatch in about twenty days.

The number of generations a year is not known, but that there are probably two is indicated by the fact that the swarms of beetles occur as a rule in June and again in August and September, but pupæ and larvæ of various sizes may be found throughout the season. The insect passes the winter both as larvæ in the soil and as beetles hidden away in sheltered places.

Treatment.

Injury from the crown girdler may be avoided in large measure by adopting the one-crop system of strawberry culture and by planting only on or near land which is not infested. As the grubs feed on the roots of various grasses, including timothy, it is not advisable to plant on land recently in sod. An immune crop, like potatoes, should intervene. The beetles may be kept from injuring the leaves of newly-set strawberry plants by thorough applications of arsenate of lead, which seems to act merely as a repellent. In cases where the plants were protected in this way the beetles have been known to enter the soil, and devour the roots, thus rendering these measures of little value.

REFERENCES

Mont. Agr. Exp. Sta. Bull. 55, pp. 130–142. 1904.
Me. Agr. Exp. Sta. Bull. 123. 1905.

BLACK VINE-WEEVIL

Otiorhynchus sulcatus Fabricius

This European weevil was apparently introduced into the Eastern states many years ago, and now occurs throughout the northern part of the country to the Pacific. The injury to the roots and crown of the strawberry are similar to that inflicted by the preceding species. The beetle is considerably larger, $\frac{3}{8}$ inch in length, and black in color; the wing-covers are marked with small patches of yellowish hairs. The insect is said to hibernate in the larval stage, the beetles appearing in April and May. In both the larval and adult conditions it sometimes becomes troublesome in greenhouses, attacking cyclamens, gloxinias, primulas, geraniums and other plants.

Its injuries may be prevented by the same measures as suggested for the crown girdler.

The Strawberry Crown-borer

Tyloderma fragariæ Riley

In the Upper Mississippi Valley strawberry plants are often seriously injured by the larva of a snout-beetle which burrows in the crown. It is a native insect and first attracted attention about 1871. The chestnut-brown beetles, about $\frac{1}{6}$ inch in length, emerge from hibernation in early spring, and the female deposits her small, elongate, yellowish-white eggs in the plant at or near the surface of the ground, singly, in cavities which she excavates with her beak. After the egg is in place, she plugs the opening with earth or bits of plant tissue. Oviposition continues until after the middle of June.

On hatching, the grub burrows downward through the crown, and by the time it is full-grown has eaten out a considerable portion of the contents. From one to three grubs may infest a plant; in the latter case only the shell of the crown is left. When full-grown the grub is about $\frac{1}{5}$ inch in length, white with a yellowish head. They become mature from the first part of July till early August, and all transform to beetles the same season. The transformation takes place within the crown, after which the beetles remain some time in the burrow in order to fully harden and then escape. At the approach of winter they go into hibernation in the soil. There is but one brood a year.

Treatment.

By adopting the one-crop system of strawberry culture and by placing new beds at some distance from infested fields, the injury caused by the crown-borer may be largely prevented. The beetles are unable to fly, and their spread from one field to another is consequently slow. If the new plants are dug in the spring before the eggs are laid there is little danger of introducing the pest into new beds unless some of the hibernating beetles are carried over in the soil adhering to the roots.

REFERENCES

Forbes, 12th Rept. St. Ent. Ill., pp. 64–75. 1883.
Ind. Agr. Exp. Sta. Bull. 33, pp. 41–43. 1890.

A closely related species (*Tyloderma foveolatum* Say), which commonly breeds in the stems of the evening primrose and willow herb, has been reported as a serious pest of the strawberry in British Columbia. The grubs attack the crown like the foregoing species.

REFERENCE

Fletcher, Rept. Ent. Bot. for 1897, p. 204. 1898.

FULLER'S ROSE BEETLE

Aramigus fulleri Horn

This well-known and destructive greenhouse pest occurs from the Atlantic to the Pacific and has been introduced into the Hawaiian Islands. In California the grubs have caused severe injury to strawberry plants and sometimes attack the roots of blackberries and loganberries.

The adult is a grayish-brown snout-beetle (Fig. 335) with an oblique whitish bar on each wing-cover. The beetles feed on the foliage of a large number of cultivated and wild plants and have been known to eat off the stems of apples, causing the fruit to drop.

The female deposits her smooth, yellow, ovoid eggs, about $\frac{1}{25}$ inch in length, in flattened clusters of ten to sixty in crevices at the base of the strawberry plant (Fig. 336). On hatching the grubs descend into the ground, where they at first devour the slender roots and later burrow into the crown, killing the plant. When full-grown the grub is about $\frac{1}{3}$ inch in length, milk white in color, and strongly arched. When mature they leave the crown and transform to pupæ in earthen cells, two to five inches below the surface of the ground. The length of time spent in the various stages has not

Fig. 336. — Egg-mass of Fuller's rose beetle. Knight photo. Enlarged.

Fig. 335. — Fuller's rose beetle. Knight photo.

been definitely determined. There are probably several broods a year. The adults sometimes riddle the leaves of citrus trees, especially in nurseries.

Treatment.

The injuries occasioned by this insect may be prevented by adopting a short rotation system of strawberry culture and by shifting the beds to new, uninfested land. The beetles have no functional wings and their natural spread to new fields is consequently slow. A careful watch should be kept of the newly set plants in April and May, and as soon as any begin to look sickly or die they should be removed and destroyed. If the work is done carefully, many of the grubs will adhere to the roots.

Experiments in California have shown that a large percentage of the grubs can be killed without injury to the plants by injecting one third ounce of carbon bisulfide into the soil every 18 inches in the rows. This should be done early in the season before any of the grubs have transformed. This treatment is too laborious and expensive for use on a large scale.

REFERENCES

Riley, Rept. U. S. Ent. for 1879, pp. 255–257. 1880.
Chittenden, Bur. Ent. Bull. 27, pp. 88–96. 1901.
Maskew, Bur. Ent. Bull. 44, pp. 46–50. 1904.
Van Dine, Haw. Agr. Exp. Sta. Press Bull. 14. 1905.

THE STRAWBERRY ROOT-WORMS

The larvæ or grubs of three species of small leaf-beetles attack the roots of the strawberry and frequently cause considerable loss, especially in old beds. While the beetles are easily separated the grubs are so much alike that it is very difficult to distinguish the different kinds and they are therefore known collectively as the strawberry root-worms. When mature they are from $\frac{1}{8}$ to $\frac{1}{6}$ inch in length, white with a pale yellowish-brown head and first segment and are strongly curved like the common white-grub. Although the injury inflicted by the various species is practically the same, they differ considerably in habits and life history.

Typophorus canellus Fabricius

This is the most abundant and destructive of the root-worms and is often referred to as the strawberry root-borer. The beetles are about $\frac{1}{8}$ inch in length and vary greatly in coloration from nearly black to reddish-yellow with blackish spots on the wing-covers; the latter vary greatly in size, shape and distinctness. The beetles hibernate under mulch or other convenient shelter and become active in early spring. They feed voraciously on strawberry leaves and show a special fondness for the opening leaves of the red raspberry, occasionally attacking blackberries (Fig. 337). The leaves of young strawberry plants are sometimes riddled

FIG. 337. — Blackberry vines injured by strawberry root-worm beetles.

and we have seen red raspberries so completely defoliated in May that the canes were killed back nearly to the ground. They are most abundant in late May and early June, at which time most of the eggs are probably laid on or near the surface of the ground near the plants. The grubs burrow through the soil, feeding on the roots, and become full-grown during July and August. When mature they construct small smooth-lined earthen cells, within which the transformation to the pupa and later to the adult takes place. The beetle remains some time in the cell in order to become fully hardened and then emerges to feed on the leaves for a time before going into hibernation at the approach of cold weather. Most of the beetles emerge during August, but a few stragglers may appear later. There is but one brood a year.

Graphops pubescens Melsheimer

This beetle is about $\frac{1}{8}$ inch in length, of a metallic coppery color and is sparsely clothed with a grayish pubescence. The beetles are most abundant and deposit eggs from June till August. The grubs become mature before cold weather and construct earthen cells in which they pass the winter in the larval condition. Pupation occurs in May and June and the beetles begin to emerge in early June. Beetles are sometimes taken as early as March, but those probably hibernated as adults. There is only one brood a year.

Colaspis brunnea Fabricius

This is also occasionally a grape pest. The grubs are sometimes found feeding on the roots of the strawberry.

Treatment for root-worms.

The losses occasioned by root-worms may be avoided in great measure by adopting a short rotation in growing strawberries. New beds should be planted on uninfested land and should be isolated from the older beds. Injury to the leaves by the beetles may be prevented by thorough spraying with arsenate of lead, 4 to 6 pounds in 100 gallons of water, as soon as the beetles appear. Of course this poison could not be used with safety after the fruit has attained much size.

REFERENCE

Forbes, 13th Rept. St. Ent. Ill., pp. 150–177. 1884.

White Grubs

Lachnosterna (several species)

White grubs are without doubt the most troublesome and destructive pests with which the strawberry grower has to contend. These large, thick-bodied, strongly curved, dirty-white grubs (Fig. 338) thrive in grass land, and their abundance is favored by a long rotation. They feed on the roots of grasses,

grains, corn, potatoes, beets and other root crops and are often destructive to nursery stock. We have seen roots of apple stock so badly eaten that the young trees could be easily lifted from the ground with the thumb and finger. Strawberries planted on infested land are sure to suffer severely.

White grubs are the larvæ of several closely related species of the common large brownish May beetles or June-bugs (Figs. 340 and 341). They belong to several species, but so far it has been impossible to distinguish them in the larval stage. The life histories of the various species are, so far as known, very similar. The parent beetles appear in vast swarms in May and June and attract attention by their habit of coming to lights. They hide in the fields during the day, but at dusk migrate in swarms to near-by trees of various kinds, where they feed on the leaves; at daybreak they return to the fields. The female burrows in the soil and deposits her eggs singly or in small groups in the

Fig. 338. — A white grub.

Fig. 339. — A white grub in its cell in the ground. Knight photo.

ground one to three inches from the surface. The soil adheres to the egg, thus forming a compact oval ball. The eggs hatch in ten to eighteen days and the grubs feed during the remainder of the season on the roots of their food-plants at an aver-

age depth of about three inches. At the approach of cold
weather they burrow deeper and hibernate at an average depth
of ten inches. A few, however, may remain near the surface,
while others may descend to a depth of two feet or more. They
return to the roots in early spring and complete their growth by
June or July. The grub then constructs an oval earthen cell
(Fig. 339) a few inches below the surface within which it soon
transforms to a delicate helpless pupa. The insect remains in
this condition till August or September and then transforms to

Fig. 340.—Two species of June-beetles,
the adults of the white grub, *Lachnosterna
ilicis* and *L. hirticula.*

Fig. 341. — A June-beetle,
Lachnosterna fusca.

the adult or beetle. As a rule, these beetles remain in the pupal
chamber till the following May or June. It is thus seen that
while the grubs mature in about two years the whole life-cycle
from egg to egg requires three years.

Treatment.

White grubs are most abundant in land which has been for
some time in sod or has been occupied by old strawberry beds.
Strawberries should never be planted on badly infested land;
owing to lack of other food, the grubs concentrate on the strawberry roots and soon kill the plants. Clover is not attacked to
any great extent by white grubs, and may be used to advantage
as an intervening crop between sod and strawberries. Much

of the loss occasioned by white grubs can be avoided by adopting the one-crop system of strawberry culture and alternating with some crop more or less immune, such as beans or peas. In the South sweet potatoes are often raised as an alternate crop. Experiments in Illinois have shown that hogs are of great value in cleaning up badly infested fields in the fall.

When newly set plants show by their lack of vigor that grubs have attacked the roots there is nothing to do but to dig out and kill the culprit by hand. This will pay in small beds, but is too expensive and laborious to be practicable in large fields.

References

Forbes, 18th Rept. State Ent. Ill., pp. 109–144. 1894.
Ill. Agr. Exp. Sta. Bull. 116. 1907.

Other Strawberry Insects

Oblique-banded leaf-roller: *apple*, p. 65.
Flea-beetles: *apple*, p. 203.
Raspberry leaf-roller: *raspberry*, p. 321.
Negro bug: *raspberry*, p. 324.
Rose chafer: *grape*, p. 397.

CHAPTER XIII

GRAPE INSECTS

THE relative importance of the insect enemies of the grape varies greatly from year to year in different regions of the country. In the eastern United States root worms are present in injurious numbers only along the shores of Lake Erie, yet Quaintance, in 1909, estimated the loss occasioned by them at over two million dollars.

The rose chafer is destructive only in sandy regions; leaf-hoppers are always present in most vineyards, but severe out-breaks alternate with extended periods of relative immunity; some whole sections are exempt from injury by the berry-moth, while in others it often occasions serious loss and the blossom midge has a very restricted range. Trellised vines are easily sprayed, and this method of controlling cape insects is prac-ticed by commercial vineyardists whenever practicable.

THE ROSE CHAFER

Macrodactylus subspinosus Fabricius

About the time grapes are in blossom vineyards located in sandy regions are often invaded by a hungry host of ungainly, long-legged, grayish-brown beetles (Fig. 342) about an half inch in length that first devour the blossom-buds and blossoms and then attack the newly set fruit and the foliage (Figs. 343 and 344). The rose chafer does not confine its attacks to the grape, but also feeds on the blossoms of the apple, pear, plum, black-berry, raspberry, strawberry and many other fruits; on the

397

rose, hence its common name, and on many wild trees, such as sumac, elder and the wild thorn. It is generally distributed from Maine to Colorado and southward to Georgia and New Mexico, but has attracted attention as a grape pest particularly in New York, New Jersey, Michigan, Ohio and Pennsylvania. It is most troublesome in sandy regions, especially in vineyards surrounded by waste grass lands, in which the larvæ find abundant food.

Fig. 342. — Male rose chafer (× 3).

The beetles emerge from the ground about the time grapes are ready to blossom, that is, about the middle of June in New York and in the latter half of May in southern New Jersey. They usually make their appearance in the vineyard suddenly and often in countless swarms. They feed at first on the blossom buds or blossoms and later attack the newly set fruit and the foliage. The destruction of the blossom causes thin scraggly clusters that are often scarcely worth picking. The berries that have been eaten into when small often show the seeds protruding from the wound later in the season. The injury to the foliage is rarely severe enough to cause serious damage although the leaves are often riddled by the beetles and have a tattered and ragged appearance. The beetles usually remain, feeding on the vines for ten days or two weeks and then migrate to other plants then in blossom.

The beetles remain mated for long periods, during which the female continues feeding. To deposit her eggs she burrows into the ground to a depth of three to six inches, sometimes alone and sometimes accompanied by the male. The smooth, white, oval eggs, about $\frac{1}{2}$ inch in length, are laid singly in little pockets in the soil (Fig. 344a), which are about two or three times as large as the egg. These egg-cells are about $\frac{1}{8}$ inch apart in the soil. The female is said to lay twelve eggs at each oviposition and

FIG. 343. — Rose chafers feeding on rose buds.

oviposits normally three times. In New York most of the eggs are deposited during the last week of June and the first half of July. The eggs hatch in two or three weeks.

The female selects light sandy land in which to oviposit; the heavier soils are never infested to any great extent. The recently hatched grubs are about $\frac{1}{8}$ inch in length, whitish with yellowish head and dark brown jaws. They can crawl quite readily on the venter and can wriggle along slowly on the back. When small the grubs can feed on decaying vegetable matter in the soil, but as they grow larger attack the roots of various grasses.

They become nearly full-grown by November, descend to a depth of about a foot and there spend the winter curled up in an oval earthen cell. The full-grown grubs (Fig. 345) resemble the common white grub of the June-beetle in form and general appearance, but are only about ⅘ inch in length. In the spring they come nearer the surface again and may resume feeding for a short time.

About the last of May or the first of June in New York most

FIG. 344. — Rose chafers feeding on newly set grapes.

of the grubs transform to pupæ in earthen cells three to six inches from the surface. The pupa (Fig. 346) is about ⅗ inch in length, yellowish-brown in color and has the remains of the larval skin adhering to the posterior end of the body. The pupal stage lasts from three to four weeks.

Means of control.

In sandy regions the rose chafer has been found one of the most stubborn vineyard pests to control. Handpicking and other mechanical methods of destroying the beetles have often been resorted to, but are too expensive and laborious for use in vineyards of any size. Various repellent substances have been tried and have proved of little value for the protection of the crop. Attempts to poison the beetles with arsenical sprays have been generally unsuccessful because the beetles dislike the poison and avoid the sprayed foliage; they swallow only a small quantity of the poison and die so slowly that the injury to the crop is not prevented. Recent work in western New York, however, has shown that if the arsenate of lead is sweetened with molasses or glucose, the beetles eat it readily and many are killed within twenty-four hours. At present this is the most feasible method of controlling this pest. The vines should be sprayed thoroughly at the first ap-

Fig. 345. — Larva of the rose chafer.

Fig. 344 *a.* — Eggs of the rose chafer in their earthen cells; egg, greatly enlarged.

Fig. 346. — Rose chafer pupæ (× 3).

pearance of the beetles, using 8 pounds of arsenate of lead to 100 gallons of water sweetened with two gallons of molasses. If the beetles are very abundant, a second application, about one week later, is sometimes necessary.

2 D

Much may be done to lessen the number of beetles by thoroughly cultivating the vineyards while the insect is in the pupal stage, the last week in May and the first half of June in New York. In regions where the pest is troublesome as much as possible of the land surrounding the vineyards should be kept in cultivated crops and the area of waste grass land should be reduced to a minimum.

REFERENCES

Harris, Insects Injurious to Vegetation, pp. 32–35.   1841.
Riley, Insect Life, II, pp. 295–302.   1890.
N. J. Agr. Exp. Sta. Bull. 82.   1891.
N. Y. (Geneva) Agr. Exp. Sta. Bull. 331, pp. 530–549.   1910.
U. S. Bur. Ent. Circ. 11, Rev. Ed.   1909.
U. S. Bur. Ent. Bull. 97, Pt. III.   1911.
Hartzell, Jour. Ec. Ent. IV, pp. 419–422.   1911.

The Vine Chafers

Anomala lucicola Fabricius ;   *A. marginata* Fabricius, and others

The beetles of this genus resemble in form their near relative, the common June beetle, but are smaller, and the wing-covers are usually ornamented with darker markings. At least three species occasionally attack the vine.

The larvæ live in the ground, living on the roots of grasses and other plants, often in company with the grubs of the rose chafer, which they closely resemble. They are, however, smoother in appearance and are of a clear, bright, straw-yellow color. The beetles emerge from the ground in spring or early summer, invade the vineyard in swarms and sometimes nearly defoliate the vines. Their ravages may be controlled by the same means as recommended for the rose chafer above.

REFERENCES

N. J. Agr. Exp. Sta. Rept. for 1892, pp. 449–450.   1893.
Lintner, 10th Rept. N. Y. State Ent., pp. 408–413.   1895.

The Grape-vine Flea-beetle

Haltica chalybea Illiger

Just as the grape buds are ready to burst in the spring they are often attacked by a small, dark, glossy, greenish-blue or steel-blue beetle a little less than a fifth of an inch in length that eats out the contents and thus destroys the future cane with its load of fruit (Fig. 347). During the past forty years this flea-beetle has caused serious injury and loss in vineyards in widely separated parts of the country. Its attacks, however, are usually confined to limited areas and are not, as a rule, sustained in one locality for a series of years. It is a native American insect and occurs from Kansas to Massachusetts and southward to Florida and New Mexico. The various species of wild grapes and the Virginia creeper were probably its original food-plants.

Fig. 347. — The grape-vine flea-beetle feeding on a young grape shoot. Enlarged.

The beetles hibernate under trash, in grass land around the edge of the vineyards and are sometimes found under loose bark at the base of the vine. They emerge from hibernation in April or in May in New York just as the grape buds are swelling.

Fig. 348. — Eggs of the grape-vine flea-beetle under a strip of loose bark (×12).

They soon find their way to the nearest vine and feed voraciously on the tender heart of the bursting buds, which they reach by gnawing through the protective covering either at the side or more commonly at the tips. The beetles do most of their feed-

Fig. 349. — Grubs of the grape-vine flea-beetle (×7).

ing during the warmer part of bright sunny days; on cold cloudy days they hide away under any convenient shelter. Each beetle may destroy several buds and thus cause a loss of fruit and foliage all out of proportion to the amount of tissue actually consumed.

After feeding for several days the females begin to deposit their elongate, oval, straw-colored eggs, $\frac{1}{30}$ inch in length, beneath the scales surrounding the buds, under the loose bark on the canes (Fig. 348), and more rarely in small clusters on the upper or under surface of the leaves. Oviposition may continue for over forty days and each female may deposit from twenty to over one hundred eggs. The time required for the eggs to hatch depends largely on the temperature; eggs laid in April while the weather is cool do not hatch very much before those laid in June. The egg stage may last from twenty to sixty days.

When newly hatched the grubs are scarcely a sixteenth of an inch long and of a very dark brown color, almost as dark as the blackish spots on the body. They soon make their way to the young leaves, where they begin to eat little irregular holes through the skin and into the soft inner tissues. They feed almost entirely on the upper surface, several grubs usually working on the same leaf (Fig. 351). Sometimes the grubs are numerous enough to riddle the foliage quite badly, but the actual injury to the vine from their work is never serious. The larvæ become full-grown in between two or three weeks and are then about one third inch in length, dark yellowish brown in color

Fig. 350. — Grubs of the grape-vine flea-beetle feeding on a blossom cluster.

and marked with regular rows of blackish spots; the head, legs, and thoracic and anal shields are black (Fig. 349). In New York the grubs are found on the leaves during June and the first half of July.

When the grubs have fed sufficiently they drop from the grape leaves, and after working their way from one half an inch to two inches into the ground, they twist and roll themselves about until a smooth cavity is formed in which a few days later the larva transforms to a saffron-yellow pupa (Fig. 352). The pupal stage lasts from a week to twelve days. The beetles

Fig. 351. — Grubs feeding on a grape leaf.

emerge from the ground during the latter part of July and after feeding for several weeks on the foliage of the grape and other plants go into hibernation in the early fall under fallen leaves and other rubbish, especially in waste fields and near-by wood-lots.

The grape-vine flea-beetle can be most easily controlled by spraying the vines with an arsenical poison between the middle of June and the middle of July. The grubs are then feeding on the upper surface of the leaves, where they can be easily reached with the spray. The poison can be applied to advantage combined with the Bordeaux mixture generally used at that season for the control of fungous diseases. In cases where summer spraying has been neglected and the beetles are present in the spring they may be poisoned by spraying the swelling buds very thoroughly with a strong arsenical mixture, or the beetles may be jarred into pans containing a little kerosene or on to sheets saturated with oil. The work of collecting the beetles must be done in the warmer part of bright sunny days while the beetles are most active, for in the cold raw weather it is hard to find them. When alarmed the beetles feign death and drop at the slightest jar. Shaking the vines or jarring

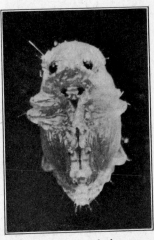

Fig. 352. — Pupa of the grape-vine flea-beetle. Enlarged.

the trellis will cause them to fall for some distance along the row. For this reason we have found it most satisfactory to collect the beetles in a small pan containing a little kerosene by tapping the canes, on which they are feeding, lightly with a small stick. As a rule the grape-vine flea-beetle is most troublesome in vineyards adjoining woodlots, hedgerows or waste land in which the beetles find suitable hibernating shelter. Wherever practicable all such conditions should be eliminated.

REFERENCES

Comstock, Rept. as U. S. Ent., pp. 213–216. 1880.
Cornell Agr. Exp. Sta. Bull. 157. 1898.
N. Y. (Geneva) Agr. Exp. Sta. Bull. 331, pp. 494–514. 1910.

The Grape Leaf-hopper

Typhlocyba comes Say

The grape leaf-hopper is our most common and widespread insect pest of the grape-vine. Practically every vineyard in the United States and Canada is infested by the insect, and

Fig. 353. — Adults of the grape leaf-hopper.

almost every year it occurs in injurious numbers in one or more localities. It has its periods of great destructiveness and of comparative obscurity, or its ups and downs, like most of our insect pests. Serious outbreaks occurred in New York in 1901 and 1902, and again in 1911 to 1912. Quayle states that with the exception of the phylloxera it is the most destructive insect pest of the vine in California. The hoppers, both adults and nymphs, suck out the sap from the under side of the leaves, which soon turn brown and, if badly infested, fall prematurely, leaving the vines unable to ripen their load of fruit or properly

Fig. 354. — Adults of the grape leaf-hopper (×3½).

mature the new wood for next year's crop.

The grape leaf-hopper has been studied most carefully in New York and in California. The adult hopper (Fig. 355) is scarcely an eighth of an inch in length, and has the back and wings marked in a peculiar manner with yellow and red. In the winter these markings are a dark orange-red, but after feeding has been resumed for a short time in the spring, they change to a light lemon-yellow. In all parts of the country it passes the winter in the adult or winged state. As soon as the leaves begin to die and drop from the

Fig. 355. — Adult and molted nymph skin of the grape leaf-hopper (× 11).

vines in September and October, the active adult hoppers migrate from the vines and seek winter quarters. Comparatively few of them find suitable hibernating places near the vines in the vineyards, unless neglect has allowed a carpet of weeds and grasses to grow. A neighboring tract of woodland or a little valley, ditch or swale overgrown with brush, or a border-

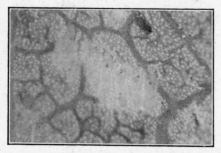

FIG. 356. — A group of eggs of the grape leaf-hopper. Greatly enlarged.

ing strip of thick grass form an ideal place for hibernation. In New York the hoppers do not feed during the winter, although on sunny days during warm spells they are sometimes seen flying about. In the milder climate of California they feed considerably during the warmer winter days on the leaves of various low-growing plants, such as alfilaria and burr-clover, in the shelter of which they find protection.

In New York the adults emerge about May 1, or a little earlier if it is warm, and begin feeding on almost any succulent growth that is available, such as young beeches and maples in the woods, and the grasses, strawberries, dewberries or raspberries near vineyards. They migrate to the

FIG. 357. — Nymphs and molted nymph skins of the grape leaf-hopper. Enlarged.

vines about the middle of May, or as soon as the grape leaves are well expanded. The lower leaves are first attacked, and as the season advances and more foliage appears, the hoppers work upward on the vines. They live almost entirely on the under side of the leaves (Fig. 353), from which they suck their food by means of their sharp beaks. They are very active creatures, leaping or flying quickly when a leaf or vine is disturbed.

FIG. 358. — Fifth stage nymph of the grape leaf-hopper.

The adults feed on the vine for two weeks or more before they begin to lay eggs. The female hopper is provided with a slender and sharp ovipositor, by means of which she inserts the eggs just beneath the lower skin of the leaf. The eggs are semitransparent, slightly bean-shaped, and about three hundredths of an inch in length and a third as wide (Fig. 356). Egg-laying continues for two months or more, and each female may deposit over a hundred eggs. In New York the first eggs are laid about the first of June in normal years and towards the middle of the month in backward seasons. The greatest number of eggs are to be found in the leaves in late June.

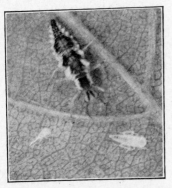

FIG. 359. — Young aphis-lion about to devour a nymph. Enlarged.

In New York the eggs of the overwintering hoppers hatch in 9 to 14 days, while in California it requires 17 to 20 days for the same brood and only 8 to 12 days for the eggs of the second or spring brood. The length of the egg stage of the second brood has not been determined in the East.

The nymphs, which are of a whitish color with red eyes when born, grow slowly and gradually acquire yellowish stripes along their sides, passing through five stages before acquiring wings (Fig. 357). In New York from 30 to 33 days are required for the development of the nymph, the first-hatched nymphs transforming to adults between the first and the middle of July, at which time the maximum number of nymphs in all stages are to be found on the leaves. In unusually long and warm seasons there may be two broods in New York, but as a rule the second brood is only partial. In California there are normally two full broods, which overlap to a considerable extent. The nymphs live on the under side of the grape leaves and suck their food like their parents, the adult hoppers. They first appear on the lower and older foliage, where the adults begin feeding, and later they infest nearly every leaf. The nymphs are active little creatures, scurrying

FIG. 360. — Grape leaves injured by leaf-hoppers late in the season.

away in all directions when disturbed, and often run to the under side of the leaves or along the leaf-stalks as if to hide. They are good runners, but, unlike the adults, are unable to jump. The effect of the work of the nymphs on the leaves is well shown in Figure 361. The upper surface of the green leaf presents a mottled, yellowish appearance. Even on badly infested vines, we have never seen the leaves so seriously injured by the nymphs in summer as by the adult hoppers either in the spring or late summer and autumn. Rarely does a leaf turn brown and die from the work of the nymphs, but the foliage is weakened so much that it falls an easy prey to the adult hoppers, into which the nymphs soon develop.

From August until the grape leaves fall, the adult hoppers swarm over the vines, sucking the life from the leaves and smutting or staining the fruit with their excrement. The loss of the foliage prevents the proper ripening of the fruit, which in consequence never acquires its proper color, flavor or sweetness.

The devastated parts of vineyards are conspicuous from a distance as brown blots on the landscape during late summer and autumn. As the leaves are killed, the hoppers migrate to other vineyards where the pasturage is better. They rise from the vines in swarms when disturbed, and often cause the pickers annoyance by getting into their eyes and ears. With the first

FIG. 361. — Grape leaf showing the work of the grape leaf-hopper nymphs.

heavy frosts the adult hoppers leave the vines and seek winter quarters in near-by grass lands or woodlots.

Means of control.

As a rule, in the East the only really serious injury to the crop is the result of the feeding of the nymphs and adult hoppers in late summer and early fall. If this can be prevented, most of the loss can be avoided. Commercial growers have, therefore, as a rule concentrated their efforts on killing as large a proportion of the nymphs as possible in early July, when the maximum number is present on the foliage and before many have transformed to adults. The nymphs are then easily killed by contact insecticides, the only difficulty being in hitting them on the under side of the leaves with the spray. At present the most satisfactory results have been attained by thorough spraying with "Black Leaf 40" tobacco extract, 1 part to 1000 parts of water. This mixture will kill all the nymphs thoroughly wet by the spray. For efficient work it is necessary to use a pressure of 125 to 150 pounds per square inch, and care must be taken to hit the under side of every leaf on which the nymphs are feeding. A fairly coarse spray is desirable, and nozzles of the cyclone type having disks of large aperture have given the best results; a fine mist spray will not wet the nymphs thoroughly enough to do good work. Sprayers fitted with fixed nozzles as are commonly used for vineyard work cannot be used against the grape leaf-hopper, as it is impossible to arrange the nozzles so as to hit the under side of the leaves with the spray. Efficient work, however, can be done by having one or two men follow the sprayer on foot, carrying a four-foot extension rod at the end of which one or two nozzles are so attached as to be directed upward. The extension rod is connected with the pump by a piece of hose 15 or 20 feet in length. By driving slowly and by having careful men to do the work thoroughly, satisfactory results in the destruction of the nymphs and the protection of the crop can be attained by one properly timed spraying.

The Geneva Agricultural Experiment Station has recently perfected a method of spraying for the grape leaf-hopper, by which the necessity of having men to carry the nozzles is dispensed with. Here the nozzles are mounted on swinging booms which are hinged to more or less rectangular frames of $\frac{3}{4}$ inch iron pipe attached to the sides of an ordinary vineyard sprayer. There are three booms on each side of the sprayer, each carrying a nozzle near the end arranged to throw the spray upward.

A spring holds the boom pressed away from the sprayer so that the nozzle is thrown into or under the vines as the sprayer is moved along the row, but can swing back when any obstacle is encountered. The exact arrangement of the nozzles will depend on the age and size of the vines as well as on the style of trimming. This apparatus is well adapted for doing efficient work in fairly level vineyards, but on steep hillsides, where it is impossible to keep the sprayer from swinging down close to the lower row, the use of trailing hose and extensions as described above will be found more practicable.

Much good may be accomplished, in preventing leaf-hopper injury, by destroying their hibernating quarters in the vicinity of vineyards. Grassy fence corners and overgrown hedge rows, in which fallen grape leaves are collected by the wind, should be removed or burned over. Clean farming also helps to control other insects and should be practiced wherever possible.

As stated above, the adult hoppers begin their work on the lower part of the vine, and it is on these lower leaves that great numbers of the eggs are deposited early in the season. Many of these lower canes or suckers are usually removed in June. If this operation could be delayed until just before spraying time, the first week in July, great numbers of the eggs would be destroyed.

In California, where the adult hoppers are more destructive in the spring and where many of the vineyards are not trellised, it has been found practicable to capture the adults in mosquito-screen cages smeared on the inside with kerosene or crude oil. The cages are made large enough to cover a vine; one side is left open, but the bottom is covered with a screen, a slit being left for the base of the vine. The cages are operated with the open side facing the wind; when the cage is placed in position, the vine receives a jar, and the hoppers, trying to escape, are caught on the oily screen. It is said that from 85 to 95 per cent of the hoppers can be captured by this method.

References

Harris, Insects Injurious to Vegetation, pp. 183–185. 1841.
Cornell Agr. Exp. Sta. Bull. 215. 1904.
Cal. Agr. Exp. Sta. Bull. 198. 1908.
N. Y. (Geneva) Agr. Exp. Sta. Bull. 331, pp. 568–579. 1910.
U. S. Bur. Ent. Bull. 97, Pt. I. 1911.
N. Y. (Geneva) Agr. Exp. Sta. Bull. 344. 1912.
U. S. Bur. Ent. Bull. 116, Pt. I. 1912.
Bull. U. S. Dept. Agr., 19. 1914.

A closely related species (*Dicraneura cockerelli* Gillette) has been known to attack the vine in New Mexico.

The Grape-leaf Skeletonizer

Harrisina americana Guérin-Méneville

More often feeding on vines grown in the garden, this native American caterpillar rarely becomes of economic importance in commercial vineyards. It is generally distributed throughout the eastern United States from New England to Florida, and westward to Missouri and Arizona, its range extending into Mexico. Its original food-plants were the Virginia creeper and wild grapes. The moths appear on the vines in the spring soon after the leaves are fully expanded. They are of a uniform blue-black color with a yellow collar, and have an expanse of about one inch. The female deposits her small lemon-colored oval eggs in loose clusters of one hundred or more on the under side of the leaves; they hatch in a week or ten days. The small yellowish white larvæ feed on the epidermis of the leaf; they remain in colonies and are usually found feeding in rows like soldiers in line. Until the larvæ reach the fifth instar, only the epidermis is consumed. They then eat holes through the leaves, devouring all the tissue except the larger veins. The full-grown larva is about one half inch in length, sulfur-yellow in color, marked with four rows of black spots visible from above; the body is clothed with rather long, bristly hairs. The caterpillars attain their growth in about forty days, descend to the ground, where they spin a tough white oval cocoon within which pupation soon takes place. Some of the pupæ may give rise to a second brood of moths in about two weeks, while the remainder do not transform until the following spring. In the latitude of Virginia there are one full and a partial second generation annually.

When infesting a few vines in the garden, it is an easy matter to destroy, by hand, the conspicuous colonies of caterpillars. In larger vineyards they may be poisoned with arsenate of lead or Paris green at the usual strength.

REFERENCE

U. S. Bur. Ent. Bull. 68, Pt. VIII. 1909.

THE GRAPE-VINE SAWFLY

Erythraspides pygmœa Say

Groups of greenish-yellow, black-spotted sawfly larvæ are sometimes found on the leaves of the vine, feeding in rows like the caterpillars of the grape-leaf skeletonizer described above. They eat the whole tissue of the leaf, however, beginning at the edge and working toward the center. The adults are four-winged flies about $\frac{1}{3}$ of an inch in length, black, with the thorax reddish above. The female deposits her eggs in small clusters on the under side of the terminal leaves. The larvæ feed in colonies, and as soon as one leaf is devoured, attack the next one below. The mature larva is a little over $\frac{1}{2}$ inch in length, greenish-yellow in color, with the head and tip of the body black; each segment has two transverse rows of black spots. When mature the larvæ enter the ground a short distance, where pupa-

2 E

tion takes place within a silk-lined earthen cocoon. There are two broods annually, the flies of the second generation appearing in late July or early August.

Like the preceding species, the grape sawfly is rarely abundant enough to seriously injure the vines. The larva may be killed with arsenate of lead or Paris green at the usual strength.

The Grape Leaf-folder

Desmia funeralis Hübner

The grape leaf-folder is widely distributed throughout the eastern United States and Canada and is present in small numbers nearly every year in most vineyards. It has attracted attention by its injuries more particularly from southern Illinois southward to Texas and eastward to North Carolina and Georgia; it also occurs in California.

The parent moths have an expanse of a little less than an inch; the wings are very dark brown, nearly black, with an opalescent reflection and are narrowly bordered with white; in both sexes the front wings have two white spots; there is one white spot on the hind wing of the male (Fig. 362) and two, often coalescent, in the female. The body of the female is crossed by two white bands, that of the male by only one.

The moths appear in the vineyards in June, and deposit their eggs singly or in small patches on the leaves. The young caterpillars feed on the upper surface of the leaf and soon begin to draw over a portion of the leaf with a few strands of silk, bringing the two surfaces together and forming a fold. The larva remains within the fold until full-grown, feeding on the upper epidermis of the leaf, which becomes skeletonized, turns brown and dies. The mature larva is grass-green in color and is about three fourths of an inch in length; when disturbed it attempts to escape by wriggling from its retreat, and suspends itself by a silken thread. The caterpillars become full-grown in about a month from the time of hatching, and transform within the folded leaf into a brownish pupa about one half of an inch in length. The moths emerge in a week or ten days, and lay eggs for a second brood.

There are two broods; the moths of the first appearing in June or July, and the second in August or September. The winter is passed in the pupa state in the folded leaves on the ground. The first brood is comparatively small and its work relatively inconspicuous; the second brood, however, is sometimes numerous enough to defoliate the vines more or less completely, and thus either prevents the proper ripening of the fruit or exposes it to injury from sun-scald. In a few instances caterpillars of the first brood have been

Fig. 362. — Male moth of the grape leaf-folder ($\times 2\frac{1}{4}$).

observed feeding in the blossom clusters which they had webbed together after the manner of the grape-berry moth.

In California the larva seems to have somewhat different habits than in the East. It rolls rather than folds the leaf and feeds on the free edge of the leaf inside the roll instead of eating off the upper epidermis.

Treatment.

The grape leaf-folder can be controlled by spraying the leaves with an arsenical poison at the time when the first-brood caterpillars are hatching, taking care to cover evenly the upper surface of the leaves. In small vineyards it is often practicable to go over the vines and crush by hand the larvæ or pupæ in the folded leaves.

REFERENCES

Riley, 3d Rept. Ins. Mo., pp. 61–63. 1871.
U. S. Dept. Agr. Farm. Bull. 284, p. 22. 1907.
Cal. Agr. Exp. Sta. Bull. 192, pp. 129–132. 1907.

THE EIGHT-SPOTTED FORESTER

Alypia octomaculata Fabricius

Grape-vines growing in city gardens are especially likely to be infested with the caterpillars of the eight-spotted forester moth. In commercial vineyards this insect rarely attracts attention. In New York the moths fly from the middle of May until the middle of June. They have an expanse of about $1\frac{1}{4}$ inches. The wings are velvety-black; there are two pale yellow spots on the front wings, and two white ones on the hind wings; the thorax and abdomen are black, at each side of the thorax there is a pale yellowish tuft of hairs. The front and middle legs are ornamented with tufts of orange hairs. They are day fliers and are often seen hovering over flowers on the nectar of which they feed. The full-grown caterpillar is about $1\frac{1}{3}$ inches in length; the head is yellowish, spotted with black; the second and third segments have a transverse row of black spots and lines. The first segment is pale orange in front; each segment back of the third has a broad central orange band on each side of which there are four narrow black bands; the second and third segments lack the orange band, but are marked with the narrow black bands; each segment is crossed by a row of black dots; below there is a series of white spots on the fourth to ninth segments; there is a rounded hump near the hind end of the body. The caterpillars are found on the vine from the first of June until the first of August. The caterpillar transforms to a pupa within a very slight cocoon upon or just below the surface of the ground.

In the garden they may be destroyed by hand-picking, or in larger vineyards they may be killed by spraying with arsenicals while the caterpillars are small.

REFERENCES

Riley, 6th Rept. Ins. Mo., pp. 88–90. 1874.
Lintner, 5th Rept. State Ent. N. Y., pp. 179–183. 1889.

THE ERINOSE OF THE VINE

Eriophyes vitis Landois

In central Europe, Italy and in California the leaves of the vine are often deformed by attacks of a minute mite. In infested leaves the portions between the larger veins puff up, leaving a cavity on the under side which is clothed with a dense felt-like covering. As a rule, thin-leaved varieties of grape are more subject to attack, or at least show greater evidence of the presence of the mite. In California erinose rarely causes serious injury to the vine or to the crop; it has often been confused with a fungous disease, the powdery mildew.

The mite, as usually found on the leaf, is a minute, nearly colorless, elongate, four-legged creature, about .13 mm. in length. The mites hibernate under loose strips of bark on the larger branches of the vine and in the spring migrate to the under side of the opening leaves where they puncture the epidermal cells with their sharp mandibles, thus producing abnormal thread-like outgrowths from the underlying layers of cells, known as erinea. These erinea when abundant have the appearance of dense felt, and it is in the shelter thus afforded that the mites live, lay their eggs and the young find suitable food. From time to time individuals leave the older leaves and start new colonies on the young leaves at the ends of the branches. In the fall some of the mites desert the leaves and go into hibernation on the bark of the older canes.

Control.

Where sulfuring is practiced for the control of the fungous disease known as oidium or powdery mildew, the mites rarely cause trouble. In severe infestations it is sometimes advisable to destroy the hibernating mites on the stump by pouring over it about a quart of boiling water during the dormant season.

REFERENCES

Landois, Zeitsch. f. wiss. Zool. XIV, pp. 353–364. 1864.
Mayet, Insectes de la Vigne, pp. 1–14. 1890.

THE GRAPE PLUME-MOTH

Oxyptilus periscelidactylus Fitch

The terminal leaves of the growing shoots of the vine are sometimes webbed together (Fig. 364) by a small greenish white-haired caterpillar, the larva of a beautiful plume-moth. The moths have an expanse of about seven tenths of an inch, yellowish-brown in color and marked with many dull whitish spots and streaks; the front wings are deeply cleft into two lobes and the hind wings divided into three lobes (Fig. 363). The entire wing margin is bordered with a long whitish or brownish fringe.

FIG. 363. — The grape plume-moth (× 3).

The eggs have not been observed. The caterpillars feed on the webbed leaves at the tip of the young shoots and are said to infest the blossom clusters, but have never been abundant enough to be a serious pest in commercial vineyards. They become full-grown the last of June or early in July and transform to peculiarly shaped green or brownish pupæ. The pupa is attached obliquely to a leaf or stem by the posterior end; it bears on the middle of the dorsal surface a large angulate projection. The moths emerge in about a week. It is not known whether there is a second brood or not, and whether the insect hibernates in the egg or adult stage.

No better remedy for this insect than hand-picking the caterpillars has been suggested. In commercial vineyards where such methods are impracticable, it does little or no harm.

Fig. 364. — Tip of grape shoot webbed together by a grape plume-moth caterpillar. Herrick photo.

Reference

Lintner, 12th Rept. N. Y. State Ent., pp. 218–222. 1897.

The Grape-cane Borer

Schistocerus hamatus Fabricius (*Amphicerus bicaudatus* Say)

The smaller shoots of the grape are often tunneled out and killed in the spring by a cylindrical dark brown beetle about $\frac{3}{8}$ of an inch in length. This beetle also attacks the apple, peach, pear and other fruit-trees, as well as certain ornamental shrubs and forest trees. It is also known as the apple-twig borer. While widely distributed throughout the United States and Canada east of the Rocky Mountains, it has been most troublesome to the grape in the Upper Mississippi Valley, in Iowa, Missouri, Kansas and Nebraska.

The grubs have been found burrowing in dying canes of the

grape, in Tamarix, a much cultivated ornamental shrub, in the subterranean stems of the Smilax or cat-briar in the South, and in the upturned roots of a maple. The eggs have not been described, but are probably laid on or in the bark of the smaller branches, for one observer records having traced the burrow to its beginning in such a location. The eggs are probably laid in the spring from April till June, for at that time the adults are most abundant and active. The burrows of the larva usually follow the pith and as a rule are three or four inches in length; they are packed full of the sawdust-like castings of the larva. The full-grown grub is whitish with brownish jaws, curved and about $\frac{2}{5}$ of an inch in length. The time normally spent in the larval stage has not been determined. The grub transforms to a reddish-brown pupa in a cell at the end of the burrow and the adult escapes by gnawing its way out to the surface. The adult is a dark brown, cylindrical beetle, a little less than $\frac{3}{8}$ of an inch in length, with the head drawn under the thorax so as to be invisible when viewed from above. The wing-covers are obliquely truncate behind and in the male this declivity is armed with a pair of blunt horn-like processes. Larvæ and pupæ have been found in grape-canes in winter, but in the majority of cases the transformation to the beetle takes place in the fall. The beetles usually hibernate in the larval burrow, but sometimes emerge in the fall and tunnel into the branches of various fruit-trees, where they pass the winter.

As breeding takes place only in dying or diseased branches the injury caused by the grubs is not great. The beetles, however, are capable of causing great damage, both to the grape and to other fruits by their peculiar habit of burrowing into the smaller branches, apparently for food and shelter only, as eggs are not laid in such situations.

Much may be done to prevent destructive outbreaks of the grape-cane borer by cutting out in the spring all diseased and dying canes in which breeding might occur and by burning all prunings. In case the beetles appear in the vineyard in the spring and begin their destructive work, the only recourse is to dig them out of the infested branches by hand or to capture them before they have entered the canes.

On the Pacific slope this species is replaced by the closely related *S. punctipennis* Leconte, the larvæ of which burrow in grape-canes.

REFERENCES

Kansas Agr. Exp. Sta. Bull. 3, pp. 27–36. 1888.
Hubbard, Ent. Am. IV, p. 95. 1888.
Lesne, Ann. Soc. Ent. Fr. LXVII, pp. 513–517. 1898.
U. S. Dept. Agr. Farm. Bull. 70, pp. 11–13. 1898.

THE GRAPE-CANE GALL-MAKER

Ampeloglypter sesostris Leconte

From Missouri to Ohio and West Virginia grape-canes are sometimes injured to a slight extent by the attacks of a reddish-brown weevil, the larva of which feeding in the cane just above one of the joints produces a swelling or gall from one to one and one half inches in length and about twice the diameter of the cane in thickness. The insect hibernates in the adult state under leaves or other suitable shelter. The weevils emerge from winter quarters in May and the female begins egg-laying as soon as the vines have made sufficient growth.

For oviposition she usually selects a place directly above the lowest joint which does not bear a fruit cluster. She first bores a hole with her snout in the heart of the cane, deposits an egg in it and then fills the cavity with bits of bark fiber scraped from the surface of the cane. She then makes a row of eight to fourteen similar punctures directly above the first and fills them with fiber, but does not deposit an egg in any except the first. As the gall increases in size this row of punctures produces an ugly wound which does not heal as long as the gall is inhabited.

The egg hatches in a week or ten days. The larva feeds principally on the pith, working its way up or down the cane for a considerable distance. When full-grown the larva is about two fifths inch in length, yellowish-white in color, with a brownish head and dark brown jaws. It becomes full-grown in eight or ten weeks, pupates within the gall and the beetles emerge in August.

The injuries caused by this insect are never serious; affected canes, unless broken by the wind, continue to grow and are able to ripen their fruit as well as those not infested. The insect may be killed by cutting out and burning the infested canes during July and early August, but ordinarily it would not be worth the trouble involved. In vineyards sprayed with an arsenical in Bordeaux mixture in late May and June many of the beetles are doubtless poisoned.

REFERENCES

Ohio Agr. Exp. Sta. Bull. 116. 1900.
W. Va. Agr. Exp. Sta. Bull. 119, pp. 323–329. 1909.

THE GRAPE-CANE GIRDLER

Ampeloglypter ater Leconte

This small black weevil is generally distributed throughout the Eastern and Western states, where it usually feeds on Virginia creeper. In West Virginia, however, it has recently attracted attention as a minor enemy of the grape.

The beetle emerges from hibernation in May and after inserting its egg in a puncture in a growing grape-cane girdles the branch below the egg and then usually eats off the tip of the branch and one or more of the leaves. These withered tips and leaves sometimes give the vine a decidedly ragged appearance, although the actual injury is not great.

The eggs hatch in about ten days and the white footless grub feeds on the pith, becoming full-grown in July. About this time the infested cane usually breaks off at the first joint below the egg puncture and drops to the ground. Pupation takes place in the cane and the beetles emerge in August.

This insect may be controlled by the same measures as suggested for the preceding species.

REFERENCE

W. Va. Agr. Exp. Sta. Bull. 119, pp. 330–339. 1909.

The Cottony Maple Scale

Pulvinaria vitis Linnæus

Grape-vines growing in the shade are sometimes infested by this conspicuous and curious scale insect, although it is rarely abundant enough to cause injury. The cottony maple scale is a native of Europe, probably introduced into this country in the early part of the last century. Its list of food-plants is extensive, including apple, pear, quince, plum, mulberry, osage-orange, box-elder, honey locust, elm, hack-berry and many others; in our cities it has become notorious as an enemy of the soft maple, hence its common name. When growing on different plants the scales vary greatly in size and form and the species has therefore received many scientific names. When living on the grape the mature female scale is about one fifth inch in length, brownish in color and resembles half a coffee-berry in form. Each female lays about 3000 pale yellowish, oval eggs in a large white cottony mass of waxen threads secreted by glands on the under side of the body (Figs. 365 and 366). As this flocculent mass increases in size the posterior end of the scale is raised from the bark at an angle of about 45 degrees. When the full number of eggs has been laid these cottony masses are much larger than the scales themselves and render infested branches highly conspicuous. The eggs are laid in late May or early June and hatching continues through June and July. The female dies soon after the last of the eggs are laid. The minute, young lice crawl to the under side of the leaves, where they develop into flattened, oval, scale-like objects, yellowish or greenish in color. From the smaller and narrower individuals winged males are produced in the early fall. After mating the males perish and the females migrate to

Fig. 365. — The cottony maple scale.

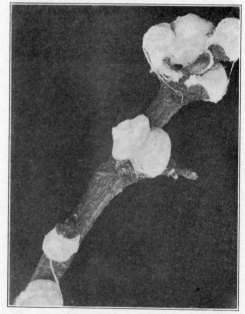

Fig. 366. — The cottony maple scale, enlarged.

the smaller branches, where they settle most often on the under side. They are then thin and flattened, but the next spring continue their growth, become strongly convex and begin egg-laying in May or June.

Means of control.

The scales and their egg-masses can often be dislodged by a stiff stream of water. Many of the young lice may be killed in summer by thorough spraying with tobacco extract and many of the over-wintering forms may be destroyed by applications of 15 per cent kerosene emulsion. The cottony maple scale is usually periodic in its occurrence, being often nearly exterminated locally by its numerous insect parasites.

REFERENCES

Réaumur, Mem. Hist. Insectes, IV, pp. 62–69. 1738.
Putnam, Proc. Davenport Acad. II, pp. 293–346. 1879.
Mayet, Insectes de la Vigne, pp. 30–36. 1890.
U. S. Bur. Ent. Bull. 22, pp. 7–16. 1900.
N. J. Agr. Exp. Sta. Rept. for 1905, pp. 591–607. 1906.
Col. Agr. Exp. Sta. Bull. 116. 1906.
Ill. Agr. Exp. Sta. Bull. 112. 1907.
Sanders, Jour. Ec. Ent. II, pp. 433–435. 1909.

THE GRAPE SCALE

Aspidiotus uvæ Comstock

This somewhat elliptical, flat, pale yellowish-brown scale measures about $\frac{1}{15}$ of an inch in diameter and has the exuvial spot, which is pale yellow with a whitish nipple, at one side of the center. It is widely distributed in the United States and occurs in Europe, Brazil and the West Indies, but is rarely injurious, attacking practically only the grape-vine, especially the crevices of bark from the ground to the second year's growth. Vines sometimes become incrusted and may die. The winter is passed in a nearly full-grown condition. The female completes her growth in the spring, and during May and June gives birth to from 35 to 50 living young. There is but a single brood annually, and the insect spreads very slowly. Ladybird beetles, parasites and mites are active enemies of this grape scale.

This pest can be controlled by thorough spraying with lime-sulfur mixture while the vines are dormant. In case this treatment has been neglected until the vines have started in the spring, the increase of the scale may be checked by several applications of either whale-oil soap, 1 pound in 4 gallons of water, or with a 10–12 per cent kerosene emulsion.

REFERENCE

U. S. Bur. Ent. Bull. 97, Pt. VII. 1912.

THE GRAPE-BERRY MOTH

Polychrosis viteana Clemens

The cause of most wormy grapes throughout the United States and Canada is the caterpillar of a small purplish-brown

moth which is always present in most vineyards and often in destructive numbers. It has been recorded in injurious numbers in Canada, Ohio, Illinois, Missouri, Pennsylvania, New York, Maryland, Virginia and Texas. In 1902 a serious outbreak occurred in the Chautauqua grape belt of western New York, where in some vineyards losses of from 25 to 50 per cent of the crop are recorded and in one case 90 per cent of the fruit was ruined.

FIG. 367. — Pupæ of the grape-berry moth. Enlarged.

The grape-berry moth always passes the winter in the pupa (Fig. 367) state in the cocoon. In the autumn most of the cocoons are found on the damp and decayed leaves close to the ground under the vines rather than on the dried leaves, which are often blown into piles, and some of the cocoons may break away from the leaves to which they are attached. The moths emerge about June 1 in New York and deposit their thin,

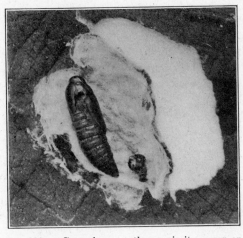

Fig. 368. — Grape-berry moth pupa in its cocoon on flap of leaf. Enlarged.

rounded, scale-like semitransparent eggs probably on the stems of the blossom clusters. Some of the caterpillars hatch and begin feeding before the grape blossoms open. They make a slight web among the blossom buds into which they eat, oftentimes destroying a dozen or more embryo grape-berries. The destructive work of this

Fig. 369. — The grape-berry moth (× 5).

spring brood of caterpillars continues in June through the blossoming period and among the recently-set berries. One

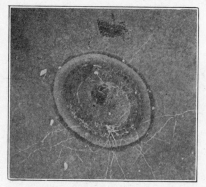

Fig. 370. — Egg of grape-berry moth on grape, greatly enlarged.

caterpillar may destroy more developing fruits in June than half a dozen caterpillars working in the larger berries later in the season. Yet this spring brood and its work rarely attracts attention because the slightly webbed portions of the clusters of blossoms or young fruits do not make infested clusters especially conspicuous (Figs. 371 and 372) and the spring brood is comparatively small.

By July 1 in New York many of the caterpillars have attained full size; they are then about three eighths of an inch

Fig. 371. — Work of spring brood of grape-berry moth caterpillars among blossoms and young fruits in June.

long and vary in color from dark greenish to purplish with a light brown head and blackish thoracic shield (Fig. 373). They go on to the leaves, where they make their peculiar cocoons, as shown in Figure 374. A little flap is cut from the leaf and gradually pulled over and down and fastened to the leaf by silken threads. The inside is then lined with white silk, thus forming a snug cocoon. At the edge of a leaf, it is necessary to cut the flap only at the ends; but when the cocoon is made away from the edge, the flap must be cut along one side also, and frequently the caterpillar cuts along where the edge of the flap is to meet the edge of the leaf and pulls up the leaf a little to meet the flap. Two to four days after building the cocoon the caterpillar transforms into a light greenish-brown pupa (Fig. 368). Many of the spring brood of caterpillars pupate during the first week in

FIG. 372. — Grape-berry moth caterpillars working among young fruits.

July and in 12 to 14 days the purplish brown moths having an expanse of a little less than half an inch begin to emerge (Fig. 369). By means of the spines on its back the pupa is

2 F

Fig. 373. — Grape-berry moth caterpillars (× 3⅓).

enabled to work its way nearly out of one end of its cocoon and the moth then emerges, leaving the empty pupa skin projecting from the cocoon. These July moths lay their eggs on the skin of the green berries or on the stems (Fig. 370).

During July and August all of the four stages of the grape-berry moth can be found in the vineyards at the same time, for the different broods overlap. The most conspicuous and destructive work is done by the second and most numerous brood of caterpillars working in the growing green grapes in July and August. This brood of caterpillars lives inside the berries on the pulp and seeds, often going into a second or third berry, fastening them together with a few silken threads (Fig.

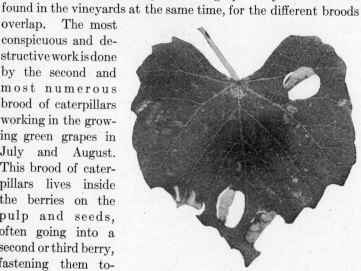

Fig. 374. — Grape leaf showing cocoons in the making and finished by grape-berry moth caterpillars.

375). Infested berries show characteristic purplish spots and often crack open, thus affording ideal places for the entrance of the spores of rot fungi (Figs. 376 and 377). Every infested berry helps to spoil the symmetry of the clusters and necessitates the labor of removing such berries before marketing the crop except where it can be sold for making the poorer grades of wine.

When the summer brood of caterpillars become full-grown in August they all go on to the leaves and cut out their characteristic cocoons. Some of the pupæ from cocoons made before

FIG. 375. — The grape-berry moth caterpillar and its work in the pulp and seeds enlarged.

the middle of August transform in 12 to 14 days into moths which lay eggs for a third or fall brood of caterpillars, but all of the second brood caterpillars pupating after that date do not emerge till the following spring. Most of the fall brood caterpillars are full-grown before October 1, but some may be found working in very ripe fruit two weeks later. A few caterpillars transform in autumn to pupæ inside the berries they infest, but most of them make their characteristic cocoons on the leaves like the earlier broods. As a rule, the insect always passes the winter in the pupa state in its cocoon on the fallen leaves.

Means of control.

The grape-berry moth can be fought most efficiently by thorough spraying with arsenate of lead, 6 pounds in 100 gallons

of water or Bordeaux mixture. The first application should be made shortly after the fruit sets; the second about ten days later and the third about the middle of July in New York, just as the eggs of the second brood are hatching. In cases of severe infestation it is sometimes necessary to make another application about ten days later. The caterpillars of the first brood do not enter the grape-berries, but feed on the outside, thus offering a better opportunity for poisoning them than is the case with the second brood. Furthermore, they are comparatively few in numbers and each one destroyed early in the season means many less caterpillars in the succeeding broods; hence the importance of thorough spraying for this brood. Thorough and careful work with a sprayer giving high pressure are necessary to throw the spray into the grape clusters, where the newly hatched caterpillars will get the poison in their first meal.

Fig. 376. — Cluster of grapes injured by the grape-berry moth-caterpillars.

Fig. 377. — Cluster of green Concord grapes badly infested by the grape-berry moth. Note the discoloration and the cracking open of the infested berries.

References

Cornell Agr. Exp. Sta. Bull. 223. 1904.
Ohio Agr. Exp. Sta. Circ. 63. 1906.

THE GRAPE-BLOSSOM MIDGE

Contarinia johnsoni Slingerland

This grape pest was first discovered in the Chautauqua grape region of western New York in 1904, and as far as known is still confined to that locality. A closely related species causes

FIG. 378. — Enlarged blossom buds infested with midge larvæ, normal blossom buds at the right.

a similar injury to grapes in Europe. Its presence in a vineyard is first indicated by abnormally enlarged, yellowish or dark reddish blossom buds (Fig. 378) which fail to open and drop off about the normal time for blossoming. Opening one of these enlarged buds, it will be found to contain a number of small whitish or yellowish maggots from $\frac{1}{16}$ to $\frac{1}{12}$ inch in length (Fig. 379). From 10 to 60 per cent of the buds are sometimes destroyed, giving the clusters a very thin and ragged appearance, and thus decreasing considerably the market value of the crop.

The parent insect is a delicate two-winged midge with a yellowish body and straw colored legs (Fig. 380). The female is about $\frac{1}{16}$

FIG. 379. — Grape blossom buds opened to show larvæ inside, enlarged.

inch in length and the male a little smaller. The flies emerge from the ground in the latter part of May just as some of the blossom buds of such early varieties as Moore, Early and Worden begin to show a small opening at the tip caused by a spreading of the petals. The female deposits her minute, grayish, elongate, slightly curved eggs in the interior of the bud by means of an extensile fleshy ovipositor inserted through this opening in the apex of the bud. From a few to seventy maggots may be found in a single bud, but twenty-five is about the average in a year of heavy infestation. The infested buds contain a watery fluid in which the maggots live. Such buds become greatly swollen, often three times as large as normal, and turn yellowish, becoming dark reddish, particularly toward the tip.

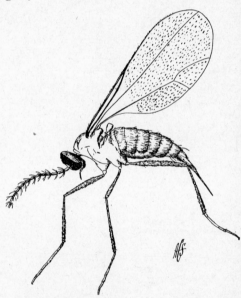

Fig. 380.— The grape-blossom midge, female (× 22).

When full-grown the maggot is about $\frac{1}{12}$ inch in length and of a yellowish or orange color. They usually escape by the opening at the apex, fall to the ground, where at a depth of about 6 inches they pass the winter as larvæ, curled up in small, ovoid, silken lined, earthen cocoons about $\frac{1}{25}$ inch in length. Pupation takes place the last of April and the adults emerge about a month later.

Control.

The grape-blossom bud midge has not been found an easy insect to control. The flies feed very little, if at all; the eggs are deposited in the interior of the bud, where maggots complete their growth out of the reach of poisons. It has been suggested that thorough cultivation of the soil might destroy a large part of the larvæ in their hibernating quarters, but the fact that many well-cultivated vineyards are badly infested would indicate that little protection can be expected from this method.

Recent experiments conducted in New York show that "Black Leaf" tobacco extract, 1 gallon in 50 gallons of water, is of considerable value in deterring the flies from depositing eggs, and thus lessens the number of injured buds. The first application should be made just as the buds of the early varieties begin to show an opening at the apex between the tips of the petals, and the second should follow in about one week.

REFFRENCES

Cornell Agr. Exp. Sta. Bull. 224, pp. 71–73. 1904.
Felt, 24th Rept. State Ent. N. Y., pp. 15–19. 1909.
N. Y. (Geneva) Agr. Exp. Sta. Bull. 331, pp. 514–530. 1910.

THE GRAPE CURCULIO

Craponius inæqualis Say

While widely distributed throughout the United States east of the Rocky Mountains the grape curculio has been most destructive in Missouri, Illinois, Ohio, Kentucky, West Virginia and North Carolina. In West Virginia losses from this cause of from 50 to 75 per cent of the crop are not uncommon. It is a native American insect which fed originally on the various species of wild grapes.

The small, inconspicuous, brownish snout-beetles, one tenth of an inch or less in length, emerge from hibernation and appear in the vineyards about the time grapes are in blossom and feed on the upper surface of the leaves for nearly a month before beginning to lay their eggs. In feeding the beetles eat only the upper epidermis of the leaf and leave characteristic feeding marks about $\frac{1}{25}$ of an inch wide and an eighth of an inch long, which soon turn whitish and give a sure indication of the curculio's presence in a vineyard.

In West Virginia egg-laying begins about the middle of June, when the grapes are about one half grown, and continues till they are ripe. The female first inserts her snout through the skin of the grape and eats out a cavity under the skin; then turning around she places a small, white, elliptical egg on the farther side of the cavity by means of her extensile ovipositor; and then seals the opening of the cavity with a drop of excrement. The egg cavity shows on the surface of the grape as a circular brownish spot with the puncture a little to one side of the center. The female may continue to lay eggs for two or three months, laying in all from 60 to nearly 400 eggs.

The eggs hatch in five to seven days, depending on the temperature, and the young grub tunnels through the fruit and usually devours one or more of the seeds. The berry often turns purplish around the egg puncture and usually drops from the vine before the grub reaches maturity, thus leaving the clusters thin and scraggled and greatly reducing the value of the crop. The grub becomes full-grown in about twelve days and then leaves the grapes through a small hole in the skin. It is white with a brownish head, legless, and at the time of leaving the grape is about $\frac{2}{7}$ of an inch in length. Pupation takes place in a small dirt-covered cocoon on or just below the surface of the ground or under the protection of a stone or piece of bark lying on the ground. The beetles emerge in about 19 days.

In West Virginia beetles of the new brood begin to appear in the latter part of July and continue to emerge until the close of the season. They are at first nearly black in ground color, but soon fade to brownish. These beetles feed on the leaves till the approach of cold weather, when they go into hibernation under rubbish, especially in near-by woodlots. Some of the earliest beetles to emerge, however, mate and lay a few eggs the same season, but as a large proportion of these eggs are infertile, only a small second brood is produced in the latitude of West Virginia.

Experiments in West Virginia have shown that the grape curculio can be successfully controlled by spraying the vines thoroughly with an arsenical poison at the first appearance of the beetles. In these experiments Paris green, one half pound in 100 gallons of water, was used, but it is probable that equally good results could be obtained by using arsenate of lead. Additional applications may sometimes be found necessary in cases where the beetles are excessively abundant.

REFERENCE

W. Va. Agr. Exp. Sta. Bull. 100. 1906.

The grape-seed chalcis (Evoxysoma vitis Saunders).

The seeds of the wild grape in the eastern United States and Canada are commonly infested by a milk-white larva about $\frac{1}{8}$ of an inch in length which eats out the kernel. The insect passes the winter as a larva within the seed. Pupation takes place in June, and in early July the adult gnaws its way out of the seed, leaving a small, round hole. The adult is a small four-winged fly about $\frac{1}{8}$ inch in length and black in color. The female is provided with a sharp needle-like ovipositor which she inserts through the skin and pulp of the grape and deposits her minute whitish egg in the kernel of the seed. Most of the egg-laying takes place in July. Cultivated varieties are rarely infested, but we have occasionally seen Delawares badly injured by the punctures made by the female in ovipositing. The injured berries color prematurely and sometimes shrivel and drop. Infestation by the grape-seed chalcis may be prevented by destroying all wild grape-vines in the vicinity of the vineyard, and by not allowing any of the infested fruit to remain in the vineyard after picking time.

REFERENCE

Cornell Agr. Exp. Sta. Bull. 265, p. 380.   1909.

The Grape Root-worm

Fidia viticida Walsh

This small grayish-brown beetle is by all odds the most destructive insect enemy of the grape occurring east of the Rocky Mountains.   It is a native American insect which doubtless fed originally on the various species of wild grapes. Its known range extends from Nebraska to Massachusetts and southward to Texas, Mississippi and North Carolina.   Since 1866 it has ranked as a pest in Kentucky, Missouri, Arkansas and Illinois, but the outbreak in the Lake Erie grape belt beginning in 1893 has been by far the most extensive, persistent and

Fig. 381. — Characteristic work of the grape root-worm beetle.

FIG. 382. — Grape root-worm beetles feeding on a leaf.

destructive in the history of this pest. In this region the insect first attracted attention in 1893 near Cleveland, Ohio;

FIG. 383. — Beetle of the grape root-worm ($\times 2\frac{1}{2}$).

in 1898 it became destructive in Erie County, Pennsylvania; in 1900 it appeared in injurious numbers in the western part of the Chautauqua grape belt of western New York and is now widely distributed throughout that region.

The grape root-worm is destructive in two stages of its development; the beetles feed on the leaves in July, eating out characteristic chain-like holes (Figs. 381 and 382) and in severe cases reducing the foliage to shreds; the grubs attack the roots, eat off the root-fibers and cut out furrows in the bark of the larger roots and main stem underground. The injury to the roots is by far the more important.

The grubs eat off all the finer roots and often strip the bark from the larger roots, thus depriving the plant of its supply of food and moisture. Badly infested vines soon take on a stunted sickly appearance,

FIG. 384. — Cluster of grape root-worm eggs, enlarged.

the leaves turn yellowish, and many fall prematurely, leaving the fruit to wither and drop. In severe cases the vines may be killed outright. Under certain conditions the insect is capable of ruining a vineyard in one or two seasons.

FIG. 385. — Grape root-worm eggs in position on grape canes.

The grape root-worm beetle (Fig. 383) is about $\frac{1}{4}$ inch in length, grayish-brown in color, with brown legs and yellowish-

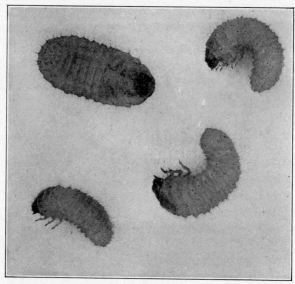

Fig. 386. — Grape root-worm, mature grubs (× 5).

brown antennæ. The beetles emerge from the ground the latter part of June and in July and at once begin feeding on the upper

Fig. 387. — Pupa of the grape root-worm.

surface of the grape leaves, producing chain-like markings, as shown in Figure 382. After feeding for about two weeks, the female deposits her first batch of eggs beneath the loose bark on the old canes and trunk of the vine (Figs. 384 and 385). Other batches are laid at intervals of about four days, each female ovipositing on an average four or five times. The total number of eggs laid by each female averages over 100. The egg is about $\frac{1}{25}$ inch in length and about one third as wide as long, cylindrical in form, with rounded ends, and is yellowish-white in color. The eggs are laid in loose, flattened clusters, averaging 25 to 40 eggs each The beetles normally begin to lay eggs the second week in July, and oviposition is at its height the latter part of the month, but eggs are deposited by a few belated individuals up to the first of September. The eggs hatch in from eight days to two weeks, depending on the temperature; in the latter part of the season incubation may last for over three weeks. Eggs laid after the first of September do not, as a rule, hatch.

On hatching the young grub drops to the ground and, taking advantage of any crack or crevice in the soil, soon burrows down to the roots of the vine. Here it feeds for the rest of the season on the finer roots, and when these are consumed attacks the larger roots and the underground part of the stem, eating out burrows and pits in the bark, as shown in Figure 389.

FIG. 388. — Grape root-worm pupa in its earthen cell.

A few of the grubs complete their growth (Fig. 386) the same season, but the majority are only about three fourths grown at the time for going into hibernation in October and November. Towards the last of May and in June the grubs ascend to within two or three inches of the surface and there construct earthen cells within which about three weeks later they transform to whitish or pinkish-white pupæ (Figs. 387 and 388). On an average, 17 days are spent in this stage. After transformation the beetles remain in the pupal cell for several days, in order to become hardened, and then burrow to the surface, emerging to-

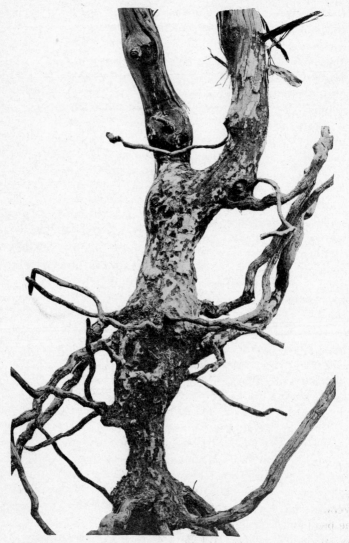

Fig. 389. — This grapevine was killed by the grubs. All of the fine feeding roots were eaten off and the main roots scarred or pitted by the grubs.

wards the last of June and in July. Normally the life-cycle is completed in one year, but under adverse conditions, such as compact clay soil or an insufficient supply of food, the larval development may be so retarded that they require a second season in which to reach maturity. In the latter case they go into hibernation early, in August of the second year, and probably give rise to the earliest beetles the following spring.

Treatment.

As the grape root-worm beetles feed extensively on the upper surface of the leaves before beginning to lay their eggs, an excellent opportunity is presented to destroy them with an arsenical spray. Arsenate of lead, 6 pounds in 100 gallons of water, is the poison now used most extensively for this purpose, replacing Paris green and arsenite of lime because of its greater adhesiveness and owing to the fact that there is less danger of foliage injury. The first application should be made very soon after the first beetles appear, for it has been shown that they eat the poison much more readily if they have never had an opportunity to feed on unsprayed foliage. A second application should be made about 10 days after the first. Recent work in western New York has shown that the efficiency of the poison can be greatly increased by adding 2 gallons of molasses to each 100 gallons of the spray liquid. Unfortunately the addition of molasses decreases the adhesiveness of the poison, and the application should not be made just before a rain, if it can be avoided. For satisfactory results in poisoning the grape root-worm, it is important that the foliage be evenly covered with a fine mist-like spray. Pumps should be used that give a high and uniform pressure, and the nozzles should be so arranged that not only the sides of the vines are hit, but the uppermost nozzle should be carried out over the top of the trellis and directed downward so as to reach the new growth, where much of the feeding is done.

As stated above, the majority of the overwintering grubs

2 G

transform to pupæ in earthen cells near the surface of the ground in the first half of June. By stirring the soil under the vines by means of a horse hoe at that time, a large part of the cells will be broken open and the tender pupæ die from exposure or fall a prey to ants or other enemies. The soil close to the base of the vine where it cannot be reached with the horse hoe should be stirred thoroughly with a hand hoe, for it is at this point that the greatest number of pupæ are found. The best results from cultivation are to be expected in years of abundant rain; in dry seasons many of the pupal cells are too deep to be reached by ordinary hoeing. By throwing up a ridge of soil under the row at the last cultivation in the summer, the grubs will be encouraged to form their pupal cells higher above the roots where it is easier to reach them with the hoe. Cultivation alone cannot be depended upon for the control of the grape root-worm, but should be employed in connection with the system of spraying suggested above.

In infested regions it is highly important to keep the vines in as thrifty a condition as possible by proper pruning, fertilization, cultivation and the use of suitable cover-crops; it is much easier to control the root-worm in a well-cared-for vineyard than in one that has been neglected and is in a run-down condition.

REFERENCES

Ohio Agr. Exp. Sta. Bull. 62. 1895.
Cornell Agr. Exp. Sta. Bull. 184. 1900.
Cornell Agr. Exp. Sta. Bull. 208. 1902.
Felt, N. Y. State Museum, Bull. 59. 1902.
Felt, N. Y. State Museum, Bull. 72. 1903.
Cornell Agr. Exp. Sta. Bull. 224, pp. 65–71. 1904.
U. S. Dept. Agr. Farmers' Bull. 284, pp. 6–12. 1907.
U. S. Bur. Ent. Bull. 89. 1910.
N. Y. (Geneva) Agr. Exp. Sta. Bull. 331, pp. 549–568. 1910.
Hartzell, Jour. Ec. Ent. IV, pp. 419–422. 1911.

The California Grape Root-worm

Adoxus obscurus Linnæus

In California the grape is attacked by a beetle which, in habits, life history and the nature of the injury inflicted, is very similar to the grape root-worm of the Eastern states. This beetle has a wide distribution in Europe, Siberia, northern Africa and the northern part of North America. It has long been recognized as a pest of the grape in France, Germany, Italy and Algeria, but in this country it has attracted attention only in California, although it is known to occur from New Hampshire to the Pacific through the Northern states and Canada. There are two forms of the beetle, a black and a brown race; in Europe only the brown form is known to attack the grape, but in California both forms are found on the vine and interbreed indiscriminately. The original wild food-plants both in this country and Europe are the various species of Epilobium or fire-weed.

The majority of the beetles emerge from the ground in the first half of May, and practically all disappear by the last of June. The beetles are about $\frac{3}{16}$ of an inch in length. There are two color varieties : one is almost entirely black; in the other the head and thorax are black, the wing-covers brown. Both varieties are clothed with a short gray pubescence. The beetles feed on the upper surface of the leaves, eating out chain-like holes or grooves; they also attack the tender shoots, and the petioles of the leaves, the fruit stems and even the small berries.

After feeding about two weeks the female begins egg-laying. The yellowish-white elongate eggs are deposited in clusters of ten to thirty in crevices beneath the strips of loose bark on the old wood. They hatch in 8 to 12 days, and the young grubs soon reach the ground and burrow down to the roots. They first feed on the smaller roots, but later attack the larger roots, eating longitudinal furrows in the bark. Some of the grubs become full-grown the first season; others complete their growth the following spring. When ready to transform, the grubs ascend to within four to eight inches of the surface, and there construct earthen cells in which the transformation to the white helpless pupa takes place. In about two weeks the pupa changes to the adult, which, after a few days required for hardening, burrows to the surface and begins its depredations on the leaves. There is only one brood a year.

Treatment.

Thorough stirring of the soil to a depth of six inches close around the vine at the time the pupæ are in their earthen cells will break many of the cells and kill a large proportion of the pupæ. By keeping the ground around the vine mulched in the spring, the moisture will be conserved, and the grubs will be more likely to come nearer to the surface for pupation. This treatment cannot be expected to fully control the root-worm, but should be used in connection with thorough spraying to kill beetles.

The beetles feed on the upper surface of the leaves for about two weeks before beginning to lay their eggs and may be killed at that time by a thorough application of arsenate of lead, 10 pounds to 100 gallons of water. When only a few vines are attacked, it is feasible to capture the beetles by jarring the vines over a canvas-covered frame or beetle-catcher.

References

Mayet, Les Insectes de la Vigne, pp. 321–332. 1890.
Cal. Agr. Exp. Sta. Bull. 195. 1908.

The Grape-vine Root-borer

Memythrus polistiformis Harris

The roots of both the wild and cultivated varieties of grapes are subject to the attacks of a large whitish borer, the larva of a beautiful wasp-like, clear-wing moth. The species ranges from Vermont to Minnesota and southward to South Carolina and Missouri. It has been most destructive in Kentucky, West Virginia and North Carolina.

The parent moths emerge in the latter part of July and usually disappear before the middle of August in the latitude of West Virginia. The female moth has an expanse of one and one half inches; the male is somewhat smaller. The front wings are opaque brown-black; the hind wings are transparent, narrowly bordered and streaked along the principal veins with violet-brown. The legs are orange, and the abdomen is crossed by two yellow bands.

The moths are most active in the heat of the day, when they may be seen flying rapidly near the ground or resting on the vines or other low vegetation. In form, color and movement they have a striking resemblance to some of the larger wasps, for which they are readily mistaken by the casual observer.

The female deposits her oval chocolate-brown eggs, about $\frac{1}{25}$ inch in length, singly or more rarely in pairs on the leaves or bark of the vines, or on the leaves or stems of grasses or weeds growing under the vines. Each female lays about 400 eggs; they are attached rather loosely, and usually fall to the ground before hatching. They hatch in about three weeks.

On hatching, the young larva at once burrows into the soil in search of a grape root on which to complete its development. After reaching the root, it burrows under the bark for a time, but as it increases in size it is able to eat out all the wood and inner bark of the smaller roots, leaving only the outer bark intact. Most of the larvæ are found a foot or more from the base of the vine. The infested roots are either greatly weakened or killed beyond the point of attack, and if many larvæ are present, the vine may suffer severely; only a small amount of bearing wood is produced, and the size of the crop is consequently small.

The larva continues to feed most of the time during the first winter and by the next fall is nearly full-grown. It spends the second winter in a silk-lined cavity in its burrow in the grape root and completes its growth the following spring. The larvæ become mature in late June and early July of the second year after hatching from the egg. They are then about one and three fourths inches in length and whitish in color, with a brownish head. When about to pupate, the larva leaves its burrow in the root, ascends almost to the surface of the ground and there constructs a silken cocoon in the outer layers of which are incorporated particles of earth and excrement. The pupal period lasts four or five weeks. When about to transform, the pupa works itself out of the cocoon so that its anterior end is brought to the surface of the ground and the moth is enabled to escape without injury to its wings. The moths usually emerge in the morning hours, the sexes mate the afternoon of the same day and the female begins to lay eggs the following day. They lay most of their eggs in about a week and die in ten days or two weeks.

Living underground in the roots at some distance from the base of the vine for the greater part of its existence, the grape-root borer may do serious injury to the vines and cause considerable loss without attracting the attention of the vineyardist. The control of this insect is not easy. Digging out the borers, as is done with its near relative, the peach-tree borer, is here impracticable, because as a rule they are found far out in the roots where it would be impossible to reach them without moving a large quantity of earth.

Much may be done, however, to lessen the numbers of moths by frequent shallow cultivation of the vineyard from the middle of June till the middle of July while the pupæ are near the surface. Not only will cultivation at that time either bury or destroy many of the pupæ, but it will tend to produce a vigorous condition of the vines and render them more resistant to borer attack.

REFERENCE

W. Va. Agr. Exp. Sta. Bull. 110. 1907.

The Grape Phylloxera

Phylloxera vastatrix Planchon

This destructive plant-louse is a native of the eastern United States, where it originally infested the leaves and roots of the various species of wild grapes. Sometime before 1863 it was introduced into the great wine-producing regions of France, and there proved a deadly enemy of the European grape (*Vitis vinifera*). By 1884 a third of the vineyards of France had been destroyed, and a much larger area had been seriously affected. It was introduced into California some time before 1874 and is now present in most of the grape-producing regions of the state except in the southern part. As most of the grapes grown commercially in California are of the *vinifera* or European type, the phylloxera has there been one of the most important insect pests with which the vineyardists of the Pacific slope have had to contend. The phylloxera has also been introduced into New Zealand and South Africa, and also occurs in southern Russia and in Algeria.

The life history of the grape phylloxera is a complicated one. In its destructive form it is a small, yellowish, wingless root-louse about $\frac{1}{25}$ of an inch in length found clustering on the roots of the vine. In feeding, the louse inserts the sharp and slender stylets of the beak and sucks out the sap, thus causing an irritation which produces an abnormal enlargement of the roots known as nodules. These swellings are yellow in color, soft and watery, and after a time break down and decay, causing the death of the root. The underground forms molt three times before becoming mature; they are all females and reproduce without being fertilized. The number of eggs laid by a single individual varies considerably with the climate and the season, but rarely exceeds a hundred. The eggs are oval in form, yellow in color and about [1] of an inch in length; they are laid singly or in clusters on the roots of the vine.

There may be six generations of the root-inhabiting form annually, and experimental rearing has shown that reproduction may continue in this way for several years without appreciable loss of vigor or of prolificness. Normally, however, during the latter half of the season some of the lice infesting the roots become more elongate than the others and have longer legs and antennæ. These forms do not become mature at the third molt and begin to lay eggs, but continue their development; at the fourth molt, wing pads become visible on the thorax; and at the fifth and last molt, which takes place after the nymph has crawled to the surface, functional wings are acquired. After waiting a few hours for the wings to dry and the body to harden, they take flight and, if a wind is blowing, may be transported a considerable distance. After feeding about a day on the under side of one of the tender leaves at the end of the shoots they are ready to deposit their yellowish-white eggs of two distinct sizes; from the larger eggs are developed the true females and from the smaller ones the true males. Both sexes are wingless, and the female after fertilization deposits a single so-called winter egg on the bark of the older canes. These sexual forms are much smaller than the others, and functional mouth parts are lacking. The insect does not pass the winter exclusively in the egg state, but many of the younger individuals of the root-inhabiting form, as well as many of the later migrants from the leaves, hibernate in small groups on the larger roots.

The winter eggs hatch in the spring soon after the first grape leaves have become well expanded, and the young lice normally crawl up the canes to the leaves, where they settle on the upper surface and begin sucking out the juices of the leaf. The puncture made in feeding stimulates the growth of the leaf cells so that a hollow gall is formed, projecting from the under side of the leaf and opening by a narrow slit on the upper surface (Fig. 390). The opening is guarded by a dense growth of down. Within this cavity the insect continues its development and becomes mature after the third molt, which occurs in about fifteen days. The forms producing the galls are all wingless females, and each individual of the earlier generation may lay from 500 to 600 eggs during the three weeks of her life as an adult. Towards the end of the season the later broods are not so prolific, 100–200 eggs being the maximum. The eggs hatch in about eight days, and the young escape from the gall

FIG. 390. — Phylloxera galls on wild grape leaf.

by the opening on the upper surface of the leaf and migrate to the tender leaves at the end of the branches, where new galls are formed. From five to seven generations of the gall-inhabiting form occurs annually. Nearly all the young of the earlier generations migrate to the leaves, but after the third generation an increasing number migrate to the roots, where they join their underground sisters in their insidious attack on the vine.

The complete life history, as outlined above, occurs only when the phylloxera is living on the species of grapes more closely related to its native food-plant. In Europe and in California the leaf galls are rarely seen. Under these circumstances it is supposed that the lice hatching from the so-called winter eggs may migrate at once to the roots. That such is the case, however, does not seem to have been proved by direct observation.

Means of control.

The grape phylloxera has been found a most stubborn pest to fight, and immense sums of money have been spent in experimental work in its control. The underground forms can be killed by injections of carbon bisulfide into the ground around the roots, provided the soil is neither too loose, so that the gas escapes too quickly, or so impervious that it does not reach the insects on the roots. The application of carbon bisulfide, however, is too expensive for use on a large scale and it is now rarely employed in commercial vineyards.

In irrigated regions it is sometimes feasible to destroy the phylloxera by inundating the vineyard for two to three months in the fall or for a longer time in the winter. The phylloxera are more easily killed by water in the summer, but unfortunately there is a great danger of injuring the vines by flooding at that time. The ground should be covered by 1 to 2 feet of water, and the submersion must be continuous for the whole period. Obviously this method of control is only of limited application.

The most practicable method of avoiding phylloxera injury, and the one most widely practiced in Europe and in California, is the use of resistant stocks. The native wild grapes of the eastern United States, while the original food-plant of the phylloxera, and usually badly infested, suffer little, if at all, from its attacks. The most resistant of these wild species which can be used for stocks on which to graft the susceptible varieties are *Vitis riparia* and *Vitis rupestris*. In the selection of resistant stocks experience has shown that many factors have to be considered if commercial success is to be attained. There are many varieties of both *Vitis riparia* and *rupestris,* many of which are worthless for stocks. A satisfactory stock must be adapted to the soil and climate of the region and be of sufficient size and vigor to give a strong, healthy growth to the vine and thus render it able to set and mature a heavy crop of high quality. The selection of resistant stocks is, therefore, a more or less local or regional problem, requiring close observation and careful experimentation by experienced growers who are thoroughly familiar with local conditions and with all the complicated problems of the grape industry.

References

Riley, 6th Rept. Ins. Mo., pp. 30–87. 1874.

Mayet, Les Insectes de la Vigne, pp. 47–147. 1890.
  A good résumé of the extensive European work on the phylloxera.

Cal. Agr. Exp. Sta. Appendix to Viticultural Rept., 1896. 1897.
  Resistant Vines.

Cal. Agr. Exp. Sta. Bull. 131. 1901.

Cal. Agr. Exp. Sta. Bull. 192, pp. 99–111. 1907.

Cal. Agr. Exp. Sta. Bull. 197, pp. 118–147. 1908.

Other Grape Insects

Climbing cutworms: *apple,* p. 138.

Twig-pruner: *apple,* p. 200.

Flea-beetles: *apple,* p. 203.

Ring-legged tree-bug: *apple,* p. 208.

Green June-beetle: *peach,* p. 296.

CHAPTER XIV

CRANBERRY INSECTS

As cranberries are usually grown in bogs which are regularly inundated during the winter and which can be reflowed at will, many cranberry pests can be controlled by a proper manipulation of the water. In some cases, however, spraying must be resorted to. As it is difficult to drive on the bogs with a sprayer, some growers have installed permanent pipes running through the beds and connected with a central power pump. The pipes are provided with outlets at regular intervals to which a rubber hose can be attached. After such a spraying plant has been installed it is a simple matter to spray a bog thoroughly.

THE BLACK-HEADED CRANBERRY WORM

Eudemis vacciniana Packard

The black-headed cranberry worm, or fire-worm, as it is more often called, is one of the most troublesome pests with which the grower has to deal. As a rule, the higher, drier bogs are less subject to injury from this source than are those regularly submerged during the winter. Submergence protects the eggs from winter-killing and also probably reduces the abundance of parasitic enemies. Bogs in which there is a heavy growth of vines are also more liable to heavy infestation.

The insect passes the winter in the egg-stage. The flat, circular, disk-like, bright yellow eggs are readily found on the under surface of the leaves. They are about half the size of an ordinary pin head, and nearly a dozen are sometimes laid on a single leaf. The eggs hatch about the time the vines begin growth in the spring. The newly hatched caterpillar is pale green with a shining black head. After feeding a day or so on the under side of the old leaf it crawls to the tip of an upright, webs together the expanding leaves and feeds within the shelter thus formed. The light green color of the under side of the leaves thus exposed in these closed tips contrasts strongly with the dark green of the normal foliage and is usually the first indication of the presence of the pest. The caterpillar becomes full-grown in about three weeks; it is then dark green in color with a black head and is about half an inch in length. It transforms to a yellow-brown pupa, usually on the ground, but sometimes within the nest of webbed leaves, and the moths emerge in about a week. By this time the leaves of the webbed tips have turned brown, and, if abundant, give the vines a scorched appearance. The moths have an expanse of about $\frac{3}{8}$ of an inch; the wings are ash gray in color and the front wings are crossed by irregular bands of brownish. The first brood of moths fly during June and lay eggs which hatch in late June and early July, in Massachusetts. The second brood larvæ make comparatively larger nests than do those of the first brood and often web together several tips. They feed on the leaves, tender buds, flowers and young fruits, giving infested areas a scorched appearance, hence the name, fire-worm, by which this pest is commonly known among cranberry growers in certain regions. The caterpillars of the second brood are, as a rule, much more abundant than of the first and the damage done correspondingly greater. These caterpillars become full-grown from the middle to the last of July, pupate on the ground and give rise to a crop of moths, which deposit the winter eggs on the under side of the leaves in late July and early August. There are two full broods each year.

Control.

Efficient work can be done against either brood of cater-pillars by thorough spraying with arsenate of lead, 6 to 7 pounds in 50 gallons of water, as soon as the eggs begin to hatch. As the eggs often hatch over a period of four or five weeks, it is sometimes advisable to repeat the application in a week or ten days. Where water is available for reflowing, a large proportion of the pupæ can be destroyed by letting the water rise up among the vines without covering them and holding it there for three days. This reflowing should be done when the greatest number of pupæ are on the ground.

A rank growth of vines tends to make the control of this insect difficult and much loss may be prevented by so managing the water and drainage as to keep down the growth of vines and encourage the production of fruit.

References

N. J. Agr. Exp. Sta. Bull. K, pp. 10–15. 1890.
U. S. Dept. Agr. Farmers' Bull. 178, pp. 9–12. 1903.
Mass. Agr. Exp. Sta. Bull. 115, pp. 6–9. 1907.
Mass. Agr. Exp. Sta. Bull. 126, pp. 3–5. 1908.
Wis. Agr. Exp. Sta. Bull. 159, pp. 6–11. 1908.

The Yellow-headed Cranberry Worm

Alceris minuta Robinson

This insect has been discussed as an apple pest on page 59. As a cranberry pest it is most troublesome on dry bogs. The green-bodied yellow-headed caterpillars web together the leaves at the tips of the uprights in practically the same way as the black-headed worms. In Massachusetts there are two broods annually, and in New Jersey three.

Unlike the preceding species the yellow-headed cranberry worm hibernates in the adult state as a slaty-gray moth having an expanse of about ¾ of an inch. They emerge from their winter hiding places in the spring and lay their minute, disk-like yellow eggs on the under side of the leaves. The eggs so closely resemble the eggs of the black-headed worm that they can be distinguished only with great difficulty. The eggs laid by the overwintering moths are fresher and brighter than those of the black-head worm which have been exposed to the weather for several months. In New Jersey the overwintering females lay their eggs in late April and early May. If the bogs are flooded at that time, they usually find a few exposed vines around the dams and along the edge of the bog. The eggs hatch in ten days and the larvæ mature in late May or early June. Pupation takes place in the nest of webbed leaves and the next brood of moths fly in June. The second brood of caterpillars are, as a rule, more abundant and destructive than the first; they make larger nests and often web together six or seven uprights and runners. They are especially fond of boring into the fruits. A third brood of caterpillars feed during August and September and give rise to a brood of moths that winter over in sheltered places, under rubbish, etc.

Control.

The moths may be prevented from laying their eggs on the cranberry vines in the spring by holding the winter flowage on the bogs till about May 20 in Massachusetts, at which time most of the moths will have disappeared. In the case of dry bogs or where it is impracticable to hold the water on the bogs late in the spring the yellow-headed worm can be readily controlled by one or two thorough applications of arsenate of lead, 10 pounds in 100 gallons of water, spraying just as the eggs are hatching. In Massachusetts the spraying for the second brood should be made about the first week in July.

REFERENCES

N. J. Agr. Exp. Sta. Bull. K, pp. 15–21. 1890.
U. S. Dept. Agr. Farmers' Bull. 178, pp. 12–17. 1903.
Mass. Agr. Exp. Sta. Bull. 115, pp. 13–14. 1907.

CRANBERRY SPAN-WORM

Cleora pampinaria Guenée

Cranberry vines are usually infested to a slight extent by various kinds of span-worms or measuring worms, also known as loopers. The most important of these, and one which is sometimes abundant enough to cause serious injury locally is the present species. The caterpillars appear in the bogs in June and become full-grown in early July. They are over an inch in length, slender, smooth and vary in color from mottled pale yellowish to brown. When full-grown the caterpillars bury themselves a short distance in the sand and transform to rough brown pupæ. The pale ash gray moths emerge in a few days and lay eggs for a second brood of caterpillars, which mature in August. The moths have an expanse of one and one half inches and have the wings sprinkled with black and crossed with diagonal toothed or scalloped blackish lines.

The first brood of caterpillars are usually found working along the edge of the bogs; those of the second brood are more evenly distributed. When abundant they strip the vines of their leaves and make the bog look as though swept by fire. They may be destroyed by spraying with arsenicals, as suggested under black-headed cranberry worms.

REFERENCES

U. S. Farm. Bull. 178, p. 19.   1903.
U. S. Bur. Ent. Bull. 66, Pt. III.   1907.

Another span-worm, *Cymatophora sulphurea* Packard, sometimes becomes destructive in Massachusetts. The caterpillars of the first brood attack the cranberry vines just as they begin to put out new foliage. They not only eat the leaves, but also destroy the buds. The full-grown caterpillar is less than an inch in length and pale green in color, with a narrow cream

colored stripe running the entire length of the body just below
the spiracles, the body striped the entire length both above
and below with fine longitudinal whitish lines. The caterpillars become mature about the middle of June and transform
to pupæ just below the surface of the sand. The sulfur yellow
moths begin to appear in a few days and continue on the wing
until the middle of July. The second brood of caterpillars
becomes mature in late July and early August.

The measures for the control of this species are the same as
those recommended above.

REFERENCE

Franklin, Ent. News, XVIII, pp. 17–20.   1907.

THE CRANBERRY GALL-FLY

Cecidomyia oxycoccana Johnson

The terminal buds of the cranberry are sometimes deformed
into a gall-like growth by the presence of several yellowish or
orange-red maggots about one sixteenth inch in length. Infested buds are killed and when the injury occurs late in the
season the formation of fruit buds for the following year may
be prevented. The larva becomes full-grown in about ten
days, spins a delicate cocoon inside the deformed bud and there
transforms to a pupa. The adults emerge a few days later;
they are delicate, two-winged, mosquito-like flies, having an
expanse of less than an eighth of an inch. The females have
the abdomen deep red and the sides of the thorax yellowish;
the male is a uniform gray. The female is provided with an
extensile ovipositor at the tip of the abdomen by means of
which she is able to place her minute white eggs in the center
of a developing bud. There are several broods annually.
The insect also infests loose-strife and certain heaths.

The cranberry gall-fly has never been a serious pest. It

2 H

cannot be controlled by poisons or by reflowing the bogs. In some cases it might be worth while to destroy all plants of loose-strife or heaths growing near the cranberries in which the flies might breed.

REFERENCE

U. S. Dept. Agr. Farmers' Bull. 178, pp. 17–19. 1903.

THE CRANBERRY FRUIT-WORM

Mineola vaccinii Riley

The cranberry fruit-worm is usually present in most bogs and often causes serious loss, especially in the higher, drier bogs which are not submerged during the winter.

The parent moths have an expanse of about $\frac{3}{4}$ inch; the front wings are ash-gray, mottled with black and white; the hind wings a uniform smoky gray. The moths fly during July and deposit their thin, flat, nearly circular, scale-like, pale yellowish eggs on the berries, most often at the calyx end. The eggs hatch in about five days and after feeding on the outside for a day or two the young caterpillar enters the berry through a small hole usually near the stem, which it closes with a web of silk. The caterpillar eats out the seed cavity and pulp of the berry and then migrates to a second and sometimes to a third or fourth berry before it becomes mature in late August or in September. The injured berries color prematurely, wither and drop from the vines. When full-grown the larva, which is then about $\frac{7}{8}$ inch in length and of a pale green color, descends to the ground and there just below the surface constructs a silk-lined, sand-covered cocoon, within which it remains in the larval state throughout the winter. In dry bogs pupation may occur as early as April, but where winter submergence is practiced it does not, as a rule, take place till after the water is drawn off in May. The moths emerge in July.

Control.

While the caterpillars in their winter cocoons are able to survive ordinary winter submergence of the bogs many may be killed by flowing the bogs for ten days or two weeks directly after picking. The water should then be drawn off to allow the vines to ripen. Holding the winter flowage on the bogs till the middle of May is of great value in controlling the fruitworm, but as it is likely to reduce the size of the crop it is not advisable to practice it except every third or fourth year. On dry bogs recourse must be had to spraying, although it has not as yet given fully satisfactory results; 1 pound Paris green and 2 pounds resin fish-oil soap in 50 gallons Bordeaux mixture have been recommended; the resin fish-oil soap is used as a sticker.

Many infested berries are picked with the crop and taken to the screen house, where they are screened out before the fruit is sent to market. These screenings should be burned promptly before the caterpillars have time to emerge, and after the picking season is over all cracks and crevices about the screen house should be cleaned out to destroy all the caterpillars that have spun up in them.

REFERENCES

Mass. Agr. Exp. Sta. Bull. 115, pp. 3–6.   1907.
Mass. Agr. Exp. Sta. Bull. 126, pp. 1–3.   1908.
Wis. Agr. Exp. Sta. Bull. 159, pp. 19–20.   1908.

THE CRANBERRY KATYDID

Scudderia texensis Saussure

In New Jersey this large, green, broad-winged katydid has sometimes been very destructive to the fruit, the seeds of which they devour. The adult katydid is about $1\frac{1}{4}$ inches in length and the female is provided with a flat, sickle-shaped ovipositor by means of which she inserts her flat, slightly kidney-shaped, yellowish-brown eggs in the edges of the leaves of certain grasses growing in the bogs or on the dikes and dams. Deer grass (*Panicum dichotomum*) and double-seeded millet (*P. viscidum*) are preferred for oviposition. There is only one brood a year, the winter being passed in the egg-stage. The younger katydid nymphs do not attack the berries. The berry-feeding habit does not develop until they reach the last nymphal stage.

The number of katydids can be greatly decreased by keeping the bogs free from the grasses in which they lay their eggs. On the dams where it is desirable that the grasses be allowed to grow in order to protect the banks, the tops containing the eggs may be burned off with a gasoline torch in the fall.

Various species of long-horned or meadow grasshoppers are usually abundant in cranberry bogs. They, too, are fond of cranberry seeds and aid the katydids in their destructive work. Their ravages may be prevented by keeping the bogs free from grass; clean bogs are rarely subject to attack.

REFERENCE

U. S. Farm. Bull. 178, pp. 26–30. 1903.

The Cranberry Girdler

Crambus hortuellus Hübner

In cranberry bogs along the Atlantic coast spots of considerable size are sometimes seen where the vines have been killed and have turned brown as if scorched by fire. The cause of the trouble is a sooty grayish caterpillar, about five eighths of an inch in length when mature, that lives in a flimsy silken tube at or just below the surface of the sand and feeds on the bark of the prostrate stems of the vines.

The parent moth (Fig. 391) has an expanse of about $\frac{7}{8}$ inch; the front wings are pale straw color marked with bands of yellow and silver towards the apex. The wings are folded closely around the body when at rest. In Massachusetts the moths fly from early June till late July and appear a little earlier in New Jersey. The female deposits her creamy white, oval eggs, 0.4 mm. in length, at the base of the food-plant; before hatching they turn to a pinkish-red color. One female was observed to lay 700 eggs in confinement. The eggs hatch in seven to ten days and the young caterpillars soon begin to construct the silken tube in which they live alongside the food-plant just at the surface of the ground. The larvæ feed on grass and sheep sorrel and probably other plants as well as on the cranberry. Most of the caterpillars become full-grown by November and hibernate in the larval condition inside a tight waterproof silken cocoon to which considerable sand adheres. Some of the belated caterpillars are said to feed again for a short time in the spring, but the majority transform the following May or June to a pale honey yellow pupa, about one third inch in length, without leaving the winter cocoon. The moths emerge very irregularly during June and July.

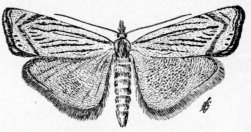

FIG. 391. — Moth of the cranberry girdler.

Control.

The cranberry girdler, working as it does at the surface of the sand beneath the layer of dead leaves and trash always found in an old bog, cannot be reached with an arsenical spray. After the larva has completed its winter cocoon it can withstand submergence for several months, but as the cocoon is not formed till November it is possible to destroy great numbers by reflowing the bogs for a week or ten days immediately after picking. Where water is not available for reflowing, badly infested areas should be burned in the spring before growth starts to kill the caterpillars, for the vines are ruined anyway and the land may then be replanted at once. The burning can be done most conveniently and safely with a gasoline torch when the vines are damp so there is no danger of the fire spreading. Bogs which are kept well sanded are rarely injured by the girdler.

REFERENCES

Cornell Agr. Exp. Sta. Bull. 64, pp. 75–76. 1894.
Scudder, Ins. Life, VII, pp. 1–5. 1894.
U. S. Dept. Agr. Farmers' Bull. 178, pp. 21–24. 1903.
Mass. Agr. Exp. Sta. Bull. 115, pp. 14–15. 1907.

THE FALSE ARMY-WORM

Calocampa nupera Lintner

Cranberry vines are often seriously injured by the caterpillar of a curiously marked noctuid moth which has an expanse of about $1\frac{3}{4}$ inches. On the fore wings there is near the center a conspicuous black spot and a V-shaped brownish spot; the apex is yellowish-brown and the hind half is purplish towards the base; the rest of the wing is streaked and banded with white and various shades of brown. The upper part of the head and the front of the thorax are covered with a dense mass of yellowish-brown scales contrasting with the reddish-brown tufts of the thorax. The moths appear in August and September and probably go into hibernation, for we have records of the moths flying in April. The eggs are laid in clusters on the under side of the leaves or on the twigs. The eggs are nearly round, slightly flattened, brownish-gray in color and marked with a series of ridges radiating from the tip. The eggs hatch in late April or early May and the young, light yellow or greenish caterpillars begin feeding on the cranberry buds, often causing a serious loss of blossoms. In the younger caterpillars the two front pairs of prolegs are not provided with hooklets and do not function in walking; these caterpillars therefore have a looping gait like a measuring-worm. As the caterpillars become larger they often defoliate the cranberry vines; they also attack various weeds and grasses. We have reared them on wild cherry leaves. When full-grown, the caterpillar is nearly two inches in length; the head is a uniform brown without

Fig. 392. — The cranberry fulgorid, adult female (× 8).

spots; the body varies in color from light green to very dark velvety brown and is striped with narrow lines of yellowish. They become full-grown in July and go into the ground, where they transform to a brownish pupa; the moths emerge in August and September.

Fig. 393. — Fifth stage nymph of the cranberry fulgorid (× 9).

Means of control.

The younger caterpillars can be poisoned by thorough applications of arsenate of lead, 4 to 7 pounds in 100 gallons of water. The older caterpillars are not easily killed by arsenicals. Where water is available the young caterpillars can be destroyed by reflowing the bogs for twenty-four to thirty-six hours soon after the middle of May in Massachusetts. If reflowing is deferred till the caterpillars are larger, many of them will be washed ashore alive and resume their destructive work when the water is drawn off. In some cases where water is not available for more than one reflow it is advisable to hold the winter flowage on the bogs till about the middle of May, thus preventing either the deposition or hatching of the eggs.

REFERENCE

Mass. Agr. Exp. Sta. Bull. 115, pp. 9–13. 1907.

THE CRANBERRY FULGORID

Phylloscelis atra Germar

Cranberry vines in the bogs of Long Island are occasionally injured by a small, broad-bodied, nearly black, jumping in-

FIG. 394. — Egg of the cranberry fulgorid. Matheson del. (× 50).

sect (Figs. 392 and 393) which in feeding, punctures the vines, causing the leaves to turn brown. The fruit does not mature properly but shrivels and drops prematurely (Fig. 395). The young insects first become noticeable in the bogs about the time of blossoming. They feed close to the ground under the protection of the vines. They are very active when disturbed and difficult to catch. They do not become mature until the last of August and in September. The adults live for a consider-able time but die off gradually at the approach of cold weather. The egg measures .8 mm. in length and is shown in Figure 394.

Control.

Experiments on Long Island would indicate that the young nymphs can be killed by a thorough application of " Black

Leaf 40″ tobacco extract, 1 pint in 100 gallons of water, adding 4 or 5 pounds of soap to make it stick and spread better.

Fig. 395. — Injured and uninjured cranberries.

CHAPTER XV

INSECTICIDES

ARSENIC in its various compounds is the cheapest and most efficient insecticide in common use. For this purpose only compounds insoluble in water can be used, since arsenic in solution is injurious to foliage, even when present in only small quantities.

ARSENIC

White arsenic, arsenious oxid (As_2O_3), is a white powder. It is the cheapest form in which arsenic can be obtained. It is soluble in water and therefore very injurious to foliage. A cheap and efficient insecticide may, however, be prepared from it by the following methods:

For use with Bordeaux mixture only. — Sal soda, 2 pounds; water, 1 gallon; arsenic, 1 pound. Mix the white arsenic into a paste and then add the sal soda and water, and boil until dissolved. Add water to replace any that has boiled away, so that one gallon of stock solution is the result. Use one quart of this stock solution to 50 gallons of Bordeaux mixture for fruit trees. Make sure there is enough lime in the mixture to prevent the caustic action of the arsenic.

For use without Bordeaux mixture. — Sal soda, 1 pound; water, 1 gallon; white arsenic, 1 pound; quicklime, 2 pounds. Dissolve the white arsenic with the water and sal soda as above, and use this solution while hot to slake the 2 pounds of lime. Add enough water to make 2 gallons. Use 2 quarts of this stock solution in 50 gallons of water.

As there is always some danger of foliage injury from the use of these homemade arsenic compounds, and as they cannot be safely combined with the dilute lime-sulfur when used as a summer spray, they are now rarely employed in commercial orchard spraying.

PARIS GREEN

Pure Paris green, $3 \ Cu(AsO_2)_2 \cdot Cu(C_2H_3O_2)_2$, is composed of copper oxid CuO, acetic acid $HC_2H_3O_2$ and arsenious oxid As_2O_3 chemically combined as copper-aceto-arsenite as follows:

Copper oxid	31.29 per cent
Arsenious oxid	58.65 per cent
Acetic acid	10.06 per cent

The commercial grades often contain impurities and vary somewhat from the above. By the National Insecticide Law of 1910 Paris green must contain at least fifty per cent of arsenious oxid, and must not contain arsenic in water soluble form equivalent to more than three and one half per cent of arsenious oxid.

For many years Paris green has been the standard arsenical insecticide for orchard use; but owing to the danger of foliage injury on stone fruits and even on pears and apples when used freely, it has now been generally replaced by the safer and more adhesive arsenate of lead. Furthermore, Paris green cannot be safely combined with the dilute lime-sulfur when used as a summer spray, nor can it be used in the self-boiled lime-sulfur on peach. In spraying apples it is usually used at the rate of eight ounces in 100 gallons of water. The danger of foliage injury is greatly lessened by using Paris green in Bordeaux mixture, or if applied in water by adding lime twice the bulk of Paris green.

LONDON PURPLE

London purple is an arsenite of lime obtained as a by-product in the manufacture of aniline dyes. Its composition is variable, the arsenic content varying from 30 to 50 per cent. It is a finer powder than Paris green and, therefore, remains longer in suspension in water. It is used in the same way as Paris green, but owing to the presence of much soluble arsenic is likely to cause foliage injury. This can be averted by the use of lime as advised under Paris green. London purple is now little used in orchard spraying.

ARSENATE OF LEAD

Arsenate of lead was first used as an insecticide in 1893, by the Gypsy Moth Commission of Massachusetts, as a substitute for Paris green, since it had been found that the latter poison would seriously injure the foliage if applied sufficiently strong to kill the gypsy moth caterpillars. It has now almost entirely replaced Paris green and London purple for orchard work throughout the country. It adheres better to the leaves, may be used at considerably greater strength without injuring the foliage and may be combined with the dilute lime-sulfur solution when used as a summer spray. Combined with the self-boiled lime-sulfur it can be safely used on the peach.

Chemically, arsenate of lead may be either triplumbic arsenate, $Pb_3(AsO_4)_2$, or plumbic hydrogen arsenate, $PbHAsO_4$. The commercial product usually consists of a mixture of these two forms, the proportion depending on the method of manufacture employed. The triplumbic arsenate of lead is prepared by combining normal sodium arsenate (Na_3AsO_4) with either lead acetate ($PbC_2H_3O_2$) or lead nitrate ($Pb(NO_3)_2$). If any di-sodium hydrogen arsenate (Na_2HAsO_4) be present, there is then formed some of the plumbic hydrogen arsenate.

Arsenate of lead is usually sold in the form of a thick paste, but for some purposes the powdered form is preferred. Under the National Insecticide Act of 1910, arsenate of lead paste must not contain more than 50 per cent water and must contain the arsenic equivalent of at least $12\frac{1}{2}$ per cent arsenic oxid (As_2O_5). The water soluble arsenic must not exceed an equivalent of $\frac{3}{4}$ of one per cent of arsenic oxid. Some of the commercial preparations contain a larger percentage of arsenic than required by the law. In the best grades of arsenate of lead paste the chemical is in a finely divided condition, and thus when diluted for use remains in suspension for a considerable time. If the paste contains less than 50 per cent water it is likely to be lumpy and requires considerable time and labor to get it into condition for use.

Arsenate of lead is used at various strengths, depending upon the insect to be killed and on the susceptibility of the foliage to injury. Four pounds in 100 gallons can be used on the peach if combined with the self-boiled lime-sulfur; on apple four or five pounds in 100 gallons is usually sufficient, although a greater strength can be used without danger of injury to the foliage; on grapes for killing the grape root-worm beetles and the rose chafer eight to ten pounds in 100 gallons have been found necessary. The poison is more readily eaten by these beetles if sweetened by two gallons of molasses in 100 gallons, but unfortunately the addition of molasses greatly decreases the adhesiveness of the poison.

Arsenate of lead is also sold in the form of a powder. In case the powder is fine enough to remain in suspension when mixed with water it gives practically as good results as the paste form. One pound of powdered arsenate of lead is equivalent to 2 pounds of the paste form.

Homemade arsenate of lead. — Haywood and McDonnell give the following directions for making arsenate of lead. " For every pound of arsenate of lead it is desired to make, use :

FORMULA A:

	Ounces.
Sodium arsenate (65 per cent)	8
Lead acetate (sugar of lead)	22

Formula B:

	Ounces
Sodium arsenate (65 per cent)	8
Lead nitrate	18

Dissolve each salt separately in from 1 to 2 gallons of water [1] (they dissolve more readily in hot water), using wooden vessels. After solution has taken place, pour slowly about three fourths of the lead acetate or nitrate into the sodium arsenate. Mix thoroughly and test the mixture by dipping into it a strip of potassium iodid test paper [2] which will turn a bright yellow if lead is in excess. If the paper does not turn yellow, add more of the lead salt slowly, stirring constantly, and test from time to time. When the solution turns the paper yellow sufficient lead salt is present, but if it should occur that the paper does not turn yellow after all the lead salt has been added dissolve a little more and add until an excess is indicated. The great advantage of this test is that it is not necessary to filter the solution or wait for it to settle.

If the paper is not at hand, the test may be made by adding a few drops of a solution of potassium iodid, when, if lead is in excess, the instant the drops touch the solution a bright yellow compound, lead iodid, will be formed.

It is very essential that the lead salt be added in *slight excess*, but a *large excess* should be avoided.

If the material has been carefully prepared with a good grade of chemicals, it will not be necessary to filter and wash the lead arsenate formed, though it would be a safe precaution to allow the lead arsenate to settle, then decant the clear solution and discard it. Approximately 1 pound of actual lead arsenate will be obtained by using the amounts of chemicals specified,

[1] The solution of lead acetate may have a milky appearance. This will be no objection and it need not be filtered.

[2] If potassium iodid paper cannot be obtained it may be prepared by dissolving a few crystals of potassium iodid in about a tablespoonful of water and saturating filter paper or blotting paper with this solution. After the paper has dried, cut into strips and keep dry until needed.

which is equivalent to practically 2 pounds of commercial lead arsenate in the paste form. It may be made up to 50 gallons with water if a formula is being used which calls for 2 pounds of commercial lead arsenate to 50 gallons, or if a stronger application is desired add less water."

ZINC ARSENITE

Arsenite of zinc, $Zn (AsO_2)_2$, is a light fluffy powder and contains the equivalent of 40 per cent arsenious oxid. It has been used extensively on the Pacific slope as a substitute for arsenate of lead. It kills insects somewhat more quickly than the latter poison and is fairly safe on apple foliage when used with Bordeaux or lime. When used in water or sweetened with molasses or glucose it causes severe injury to grape foliage, but may be safely used with Bordeaux mixture. It is probable that foliage injury by zinc arsenite is due to the solubility of this poison in water containing a small quantity of carbonic acid; the latter is usually present on the leaves, being derived from the respiration of the plant. One pound of zinc arsenite is equivalent in effectiveness to about three pounds of arsenate of lead.

HELLEBORE

Hellebore is a light brown powder made from the roots of the white hellebore plant (*Veratrum album*), one of the lily family. It is applied both dry and in water. In the dry state, it is usually applied without dilution, although the addition of a little flour will render it more adhesive. In water, 4 ounces of the poison is mixed with 2 or 3 gallons, and an ounce of glue, or thin flour paste, is sometimes added to make it adhere. A decoction is made by using boiling water in the same proportions. Hellebore soon loses its strength, and a fresh article should always be demanded. It is much less poisonous than the arsenicals, and should be used in place of them upon ripening fruit. It is used for various leaf-eating insects, particularly for the currant worm and rose slug.

SOAPS

Soap solutions are often used as contact insecticides for killing plant-lice and other small, soft-bodied insects. The so-called whale-oil or fish-oil soaps are most widely used for this purpose. The commercial brands are usually by-products from the manufacture of other products, and contain many impurities; furthermore, many of them contain an excess of free or uncombined alkali and are consequently very likely to injure young and tender foliage. An excellent fish-oil soap may be easily prepared at home by the following formula:

Caustic soda	6 pounds
Water	½ gallon
Fish-oil	22 pounds

Completely dissolve the caustic soda in the water, and then add the fish-oil very gradually under constant and vigorous stirring. The combination occurs readily at ordinary summer temperatures and boiling is unnecessary. Stir briskly for about twenty minutes after the last of the oil has been added.

A good insecticide soap can be prepared in a similar way from cotton-seed oil soap-stock or from the more impure grade known as pancoline. In fact, there is on the market a good insecticide soap made from similar materials.

SULFUR

Sulfur is commonly sold in two forms, — flowers of sulfur and flour of sulfur. Flowers of sulfur or sublime sulfur is a fine, impalpable yellow powder insoluble in water, and is formed by condensing sulfur vapor in a large chamber of brick work. If the sulfur vapor is condensed to the liquid form in a cold receiver, roll sulfur is formed. Flour of sulfur is made by grinding roll sulfur to a fine powder.

In the form of a powder or dust, sulfur is especially valuable against red spider. In California flowers of sulfur mixed with equal parts of hydrated lime is blown on the trees for the control of red spider and mite. Sulfur is sometimes used for the same purpose mixed with water at the rate of 1 pound in 3 gallons of water, in which a little soap has been dissolved to help keep the sulfur in suspension. The sulfur settles quickly and should be agitated constantly during spraying. The sulfur will remain in suspension longer if first made into a paste with water containing $\frac{1}{2}$ of 1 per cent of glue.

Lime-sulfur Solution

A solution of lime and sulfur was first used as an insecticide in California in 1886. The mixture at that time was known as the lime-sulfur and salt wash, but experience has shown that the presence of the salt does not increase the value of the wash, and it is now usually omitted. The lime-sulfur solution is the most widely used, safest and most efficient contact insecticide now available for the control of scale insects and blister-mite.

In many regions it has also replaced Bordeaux mixture for the control of fungous diseases on apple and pear.

When lime and sulfur are boiled together in water, a complicated chemical reaction takes place. The calcium (Ca) contained in the lime (CaO) combines with the sulfur (S) in varying amounts. Two of the compounds thus formed are calcium pentasulfid (CaS_5) and calcium tetrasulfid (CaS_4), containing respectively 80 and 76 per cent of sulfur. At the same time there is always formed a smaller quantity of thiosulfate (CaS_2O_3). These three compounds are soluble in water and give to the solution its insecticidal value. It is supposed that a solution will be more effective in proportion as it contains a higher percentage

21

of the pentasulfid. To insure the complete union of the sulfur and lime, it is necessary to boil the mixture about one hour. If boiled much less than an hour, some of the ingredients will be left in a free and insoluble condition, forming a sediment. If boiled much more than an hour, insoluble compounds of calcium and sulfur are formed, which go to increase the quantity of sediment.

To avoid the formation of sediment in the solution, it is important that only high grade lime be used, that the lime and sulfur be combined in proper proportions, and that the mixture should not be cooked in too concentrated a form.

To obtain the best results, use lime guaranteed to contain at least 95 per cent calcium oxid; lime containing less than 90 per cent should be avoided. Lime containing more than five per cent magnesium oxid should never be used, as the presence of the magnesium causes an unnecessary loss of sulfur, produces the poisonous hydrogen sulfid gas (H_2S) and increases the amount of sediment. In order to avoid the presence of uncombined lime or sulfur in the mixture, twice as much sulfur as lime is used (90 per cent calcium oxid), since this is the proportion in which they combine under these conditions.

Several formulas have been used in the past, but the following, worked out at the Geneva Agricultural Experiment Station, is on the whole the most satisfactory for commercial work.

Homemade concentrated lime-sulfur solution.

Lump lime	{ 95 per cent calcium oxid	38 pounds
	{ 90 per cent calcium oxid	40 pounds
Sulfur		80 pounds
Water		50 gallons

Make a paste of the sulfur with about 10 gallons of hot water. Add the lime. As the lime slakes add hot water as necessary to prevent caking. When the lime has slaked add hot water to make 50 gallons and boil one hour, stirring constantly. Water should be added from time to time to keep the liquid up to 50 gallons. Store in air-tight hardwood barrels. Test the strength of the solution with a Beaumé hydrometer and dilute for use according to the following table:

DILUTIONS FOR DORMANT AND SUMMER SPRAYING WITH LIME–SULFUR MIXTURES

READING ON HYDROMETER	AMOUNT OF DILUTION NUMBER OF GALLONS OF WATER TO ONE GALLON OF LIME-SULFUR SOLUTION		
	For San José scale	For blister-mite	For summer spraying of apples
Degrees Beaumé			
35	9	$12\frac{1}{2}$	45
34	$8\frac{3}{4}$	12	$43\frac{1}{4}$
33	$8\frac{1}{4}$	$11\frac{1}{2}$	$41\frac{1}{2}$
32	8	11	40
31	$7\frac{1}{2}$	$10\frac{1}{2}$	$37\frac{3}{4}$
30	$7\frac{1}{4}$	10	$36\frac{1}{4}$
29	$6\frac{3}{4}$	$9\frac{1}{2}$	$34\frac{1}{4}$
28	$6\frac{1}{2}$	9	$32\frac{3}{4}$
27	6	$8\frac{1}{2}$	31
26	$5\frac{3}{4}$	8	$29\frac{1}{2}$
25	$5\frac{1}{4}$	$7\frac{1}{2}$	$27\frac{3}{4}$
24	5	7	26
23	$4\frac{1}{2}$	$6\frac{1}{2}$	$24\frac{1}{4}$
22	$4\frac{1}{4}$	6	$22\frac{3}{4}$
21	$3\frac{3}{4}$	$5\frac{1}{2}$	$21\frac{1}{4}$
20	$3\frac{1}{2}$	5	$19\frac{3}{4}$
19	$3\frac{1}{4}$	$4\frac{3}{4}$	$18\frac{1}{4}$
18	3	$4\frac{1}{4}$	17
17	$2\frac{3}{4}$	4	16
16	$2\frac{1}{2}$	$3\frac{3}{4}$	15
15	$2\frac{1}{4}$	$3\frac{1}{2}$	14
14	2	3	$12\frac{3}{4}$

An older formula and the one most widely used in the past is given below. This does not give a concentrated wash but the mixture is applied at the original strength without dilution.

Homemade lime-sulfur.

Quicklime	20 pounds
Sulfur (flour or flowers)	15 pounds
Water	50 gallons

The lime and sulfur must be thoroughly boiled. An iron kettle may be used for this purpose, or the mixture may be cooked in

Fig. 396. — A plant for cooking lime-sulfur solution.

barrels by forcing live steam into it through a pipe or rubber hose (Fig. 396). Place the lime in the kettle or barrel and add hot water gradually in sufficient quantity to produce the most rapid slaking of the lime. If too much water is added at first, it "drowns" the lime and slaking takes place very slowly. When the lime begins to slake add the sulfur and stir it in thoroughly. When the slaking is completed add more water and boil the mixture about one hour. As the lime and sulfur go into solution a rich orange-red or olive-green color will appear, depending on the kind of lime used. After boiling one hour add water to the required amount and strain into the spray tank. This wash is most effective when applied warm, but may be used cold.

Owing to the excess of lime used a solution made by the above method will contain a large quantity of sediment. As this sediment is liable to clog the nozzles and interfere with the application of the wash, and as it has little insecticidal value, this method of preparation has been abandoned by most commercial growers, who now use the concentrated solution.

Concentrated commercial lime-sulfur. — There are on the market many brands of the concentrated lime-sulfur solution. They usually test from 30 to 33 degrees Beaumé and should for use be diluted according to the table given above.

Self-boiled lime-sulfur. — This preparation is used primarily as a fungicide for the prevention of brown rot on stone fruits. This is not a boiled solution, as might be inferred from the name. It is prepared by placing in a barrel 8 pounds of the best stone lime, to which is added a small quantity of cold water in order to start it slaking. Eight pounds of sulfur worked through a sieve to break up the lumps is then added slowly to the slaking lime, which is kept from burning by the addition of just enough cold water so as not to drown it. The slaking mixture must be stirred constantly. Just as soon as the slaking is completed (which should be in five to fifteen minutes), fill the barrel with cold water (50 gallons). The mixture is strained into the sprayer tank through a sieve of 20 meshes to the inch. It must be agitated constantly while being applied, as it settles rapidly. When properly made this is simply a fine mechanical mixture of lime and sulfur produced by the heat and bubbling action of slaking, and should have but little sulfur in solution. This mixture is especially adapted for the spraying of peaches and plums in foliage, as it causes no injury. Arsenate of lead may be added to this mixture for the control of plum curculio.

EMULSIONS

Emulsions are oily or resinous sprays in which these substances are suspended in water in the form of minute globules, a condition brought about by the addition of soap. They form an important class of contact insecticides useful particularly against scale insects and plant-lice.

KEROSENE EMULSION

Kerosene emulsion is one of the oldest of our contact insecticides. It is especially valuable for use against plant-lice and other small, soft-bodied insects. It is prepared by the following formula :

Soap	$\frac{1}{2}$ pound
Water	1 gallon
Kerosene	2 gallons

Dissolve the soap in hot water ; remove from the fire and while still hot add the kerosene. Pump the liquid back into itself for five or ten minutes or until it becomes a creamy mass. If properly made the oil will not separate on cooling.

For use on dormant trees, dilute with 5 to 7 parts of water. For killing plant-lice on foliage, dilute with 10 to 15 parts of water. Crude oil emulsion is made in the same way by substituting crude oil in place of kerosene. The strength of oil emulsions is frequently indicated by the percentage of oil in the diluted liquid:

For a 10 per cent emulsion add 17 gal. water to 3 gal. stock emulsion.
For a 15 per cent emulsion add 10⅓ gal. water to 3 gal. stock emulsion.
For a 20 per cent emulsion add 7 gal. water to 3 gal. stock emulsion.
For a 25 per cent emulsion add 5 gal. water to 3 gal. stock emulsion.

Distillate Emulsion

Distillate emulsion is widely used in California.

Distillate (28° Beaumé)	20 gallons
Whale-oil soap	30 pounds
Water	12 gallons

Dissolve the whale-oil soap in the water, which should be heated to the boiling point, add the distillate and agitate thoroughly while the solution is hot. For use add 20 gallons of water to each gallon of the stock emulsion.

Carbolic Acid Emulsion

This spray is used in California for mealy bugs, plant-lice and the soft brown scale.

Whale-oil soap . . . . . . , . . .	40 pounds
Crude carbolic acid	5 gallons
Water	40 gallons

Dissolve the soap completely in hot water, add the carbolic acid, and heat to the boiling point for 20 minutes. For use add 20 gallons of water to each gallon of stock emulsion.

Miscible Oils

There are now on the market a number of concentrated oil emulsions, known as soluble or miscible oils, intended primarily for use against the San José scale. For this purpose they are fairly effective when diluted with not more than 15 parts of water. To lessen danger of injury to the trees applications should not be made when the temperature is below freezing, nor when the trees are wet with snow or rain. Trees are less susceptible to injury just before the buds start in the spring. Methods have been devised for preparing these concentrated emulsions at home, but as there is considerable danger attending the process, it is better to buy them ready-made.

Tobacco

Tobacco is one of our most useful insecticides. The poisonous principle in tobacco is an alkaloid, nicotine, which in the pure state is a colorless oily fluid, slightly heavier than water, of little smell when cold and with an exceedingly acrid burning taste even when largely diluted. It is soluble in water and entirely volatile. It is one of the most virulent poisons known; a single drop is sufficient to kill a dog.

Commercial tobacco preparations have been on the market for many years. The most important of these are Black Leaf, "Black Leaf 40" and Nicofume.

Black Leaf was formerly the most widely used tobacco extract. It contains only 2.7 per cent nicotine and has now been replaced by the more concentrated extracts. It is used for plant-lice at the rate of 1 gallon to 65 gallons of water.

"*Black Leaf* 40" is a concentrated tobacco extract containing 40 per cent nicotine sulfate. Its specific gravity is about 1.25. In this preparation the nicotine is in a non-volatile form, it having been treated with sulfuric acid to form the sulfate. "Black Leaf 40" is used at strengths varying from 1 part in 800 parts of water to 1 part in 1600 parts. It can be satisfactorily combined with other sprays, as, for instance, lime-sulfur solution, arsenate of lead and the various soap solutions. When used with water about 4 pounds of soap should be added to each 100 gallons to make the mixture spread and stick better.

Nicofume is a tobacco extract containing 40 per cent of nicotine in the volatile form. It is intended primarily for use in greenhouses. Strips of paper soaked in this preparation are smudged in greenhouses to destroy aphids.

Tobacco is also used in the form of dust for the same purpose. It is especially valuable against root-lice on asters and other plants. Tobacco extracts can be made at home by steeping tobacco stems in water, but as they vary greatly in nicotine content and are sometimes likely to injure tender foliage, it is better to buy the standardized extracts.

BORDEAUX MIXTURE

Bordeaux mixture has been for many years the most widely used fungicide, but has now been replaced by the dilute lime-sulfur solution for orchard spraying in many parts of the country, owing to the serious foliage injury and russetting of the fruit often caused by its use. In addition to its fungicidal properties it also acts as a deterrent to many insects, especially flea beetles. Either arsenate of lead or Paris green may be used in combination with it.

Bordeaux mixture is prepared by mixing a solution of copper sulfate ($CuSO_4$) and milk of lime (Calcium Hydroxide) according to the following formula:

Copper sulfate	4 pounds
Lime	4 pounds
Water	50 gallons

In some cases a weaker mixture is used, containing 3 pounds of copper sulfate and lime respectively. These formulas are often abbreviated thus: 4–4–50 and 3–3–50.

When needed in large amounts Bordeaux mixture is most

conveniently prepared by using a stock solution of copper sulfate and milk of lime, storing them in tubs on an elevated platform from which the desired quantity of each can be easily drawn off into the spray tank.

Dissolve the required quantity of copper sulfate in water in the proportion of one pound to one gallon, several hours before the solution is needed; suspend the copper sulfate crystals in a sack near the top of the water. A solution of copper sulfate is heavier than water. As soon, then, as the crystals begin to dissolve, the solution will sink, bringing water again in contact with the crystals. In this way, the crystals will dissolve much sooner than if placed in the bottom of the barrel of water. In case large quantities of stock solution are needed, two pounds of copper sulfate may be dissolved in one gallon of water.

Slake the required quantity of lime in a tub or trough. Add the water slowly at first, so that the lime crumbles into a fine powder. If small quantities of lime are used, hot water is preferred. When completely slaked or entirely powdered, add more water. When the lime has slaked sufficiently, add water to bring it to a thick milk or to a certain number of gallons. The amount required for each tank of spray mixture can be secured approximately from this stock mixture, which should not be allowed to dry out. Hydrated or prepared lime of good quality may be substituted for the stone lime. Place the required quantity in the barrel or tank and add water. No slaking is required. Do not use air-slaked lime.

Take five gallons of stock solution of copper sulfate for every fifty gallons of Bordeaux required. Pour this into the tank. Add water until the tank is about two thirds full. From the stock lime mixture add the required quantity. Stir the mixture, add water to make 50 gallons. Experiment Stations often recommend the diluting of both the copper sulfate solution and the lime mixture to one half the required amount before putting together. This is not necessary, and is often impracticable for commercial work. It is preferable to dilute the copper sulfate solution. Never pour together the concentrated stock mixtures and dilute afterward. Bordeaux mixture of other strengths as recommended is made in the same way, except that the amounts of copper sulfate and lime are varied according to the requirements.

It is not necessary to weigh the lime in making Bordeaux mixture, for a simple test can be used to determine when enough of a stock lime mixture has been added. Dissolve an ounce of yellow prussiate of potash in a pint of water and label it "poison." Cut a V-shaped slit in one side of the cork so that the liquid may be poured out in drops. Add the lime mixture to the diluted copper sulfate solution until the ferrocyanide test solution *will not turn brown* when dropped from the bottle into the mixture. It is always best to add a slight excess of lime.

Fumigation

The fumes of hydrocyanic acid gas are very destructive to insect life. Fumigation with this gas is practiced extensively in greenhouses, in citrus orchards, where the trees are inclosed in portable tents for the purpose, and for the destruction of scale insects on nursery stock.

Hydrocyanic acid gas is a deadly poison, and the greatest care should be exercised in its use. For generating the gas always use 98 to 100 per cent pure potassium cyanide, and a good grade of commercial sulfuric acid. The chemicals are always combined in the following proportion:

Potassium cyanide	1 ounce
Sulfuric acid	1 fluid ounce
Water	3 fluid ounces

Use an earthen dish, pour in the water first, then add the sulfuric acid to it. Put the required amount of cyanide into a thin paper bag, and when all is ready, drop it into the liquid and leave the room immediately.

Dormant nursery stock may be fumigated in a tight box or fumigating house made especially for the purpose. Fumigating houses are built of two thicknesses of matched boards, with building paper between, and are provided with tight-fitting doors, and with ventilators. The stock should be reasonably dry to avoid injury, and should be piled loosely in the house to permit a free circulation of the gas. Use 1 ounce of potassium cyanide to each 100 cubic feet of space, and let the fumigation continue 40 minutes to one hour.

INDEX

A

abdominalis, Hartigia, 334.
Ablerus clisiocampæ, 168, 179.
abnormis, Aphelinus, 175.
aculiferus, Leptostylus, 194.
Adoxus obscurus, 451.
ænea, Tischeria, 71.
ænescens, Magdalis, 199.
Agrilus ruficollis, 332.
　sinuatus, 230.
albida, Syneta, 205.
Alceris minuta, 59, 462.
　minuta cinderella, 60.
Aleyrodes packardi, 369.
Allorhina mutabilis, 298.
　nitida, 296.
Alsophila pometaria, 86.
alternata, Rhynchagrotis, 139.
Alypia octomaculata, 420.
americana, Harrisina, 416.
americana, Malacosoma, 112.
americana, Schizoneura, 157.
American plum borer, 253.
American raspberry beetle, 323.
Ampeloglypter ater, 426.
　sesostris, 425.
Amphicerus bicaudatus, 423.
amygdali, Pulvinaria, 264.
amygdalina, Caliroa, 288.
Anaphes gracilis, 168, 175.
Anarsia lineatella, 284.
Ancylis comptana, 361.
　nubeculana, 61.
ancylus, Aspidiotus, 179.
annuæ, Aphis, 152.
annulata, Brochymena, 208.
annulipes, Pimpla, 16.
Anomala lucicola, 402.
　marginata, 402.
Anomolon exile, 121.
antennata, Xylina, 39.
Anthonomus quadrigibbus, 35.
　signatus, 372.

antiqua, Notolophus, 105.
Antique tussock-moth, 105.
Apanteles cacœciæ, 58.
　hyphantriæ, 111.
　sp., 55.
Aphelinus abnormis, 175.
　diaspidis, 337.
　fuscipennis, 168, 175.
　mali, 157.
　mytilaspidis, 168, 175.
Aphids, on apple, 142.
Aphis annuæ, 152.
　forbesi, 382.
　mali, 147.
　malifoliæ, 149.
　persicæ-niger, 289.
　pomi, 147.
　scotti, 259.
　setariæ, 259.
　sorbi, 149.
apicalis, Labena, 197.
appendiculatus, Gymnonychus, 344.
Apple bud-aphis, 151.
　bud-borer, 184.
　bud-worm, 46.
　curculio, 35.
　flea-beetles, 204.
　fruit-miner, 26.
　insects, 9.
　leaf-aphis, 147.
　leaf-hopper, 180.
　leaf-hopper, Bird's, 183.
　leaf-sewer, 61.
　leaf-skeletonizer, 67.
　maggot, 31.
　red bugs, 28.
　weevil, 38.
　wood-stainer, 198.
Apple-tree borer, flat-headed, 194.
　borer, round-headed, 185.
　borer, spotted, 193.
　tent-caterpillar, 112.
Aramigus fulleri, 389.
Araneus displicatus, 59.
Archips argyrospila, 62.

2 κ

THE following pages contain advertisements of a
few of the Macmillan books on kindred subjects

INJURIOUS INSECTS
HOW TO RECOGNIZE AND CONTROL THEM

By WALTER C. O'KANE

Entomologist of the New Hampshire Experiment Station, and Professor
of Economic Entomology in New Hampshire College

Decorated Cloth.   414 pages.   Over 600 Photographic Illustrations
$2.00 net; postpaid, $2.17

Written out of a large scientific knowledge, but in a popular style, this book discusses concisely and yet fully the characteristics, life histories, and means of control of our common injurious insects.

The illustrations are from photographs throughout.   The idea of the author has been to picture graphically the injurious stages and the work of the various pests, so that they may easily be recognized, independently from the text.   More than 135 of the illustrations are photomicrographs.

All of the common injurious forms are described, including the pests of orchard, garden, field crops, domestic animals, and the household.

The arrangement of species is original and unique.   In each division the pests are grouped according to the place where found at work and the characteristics.   Prompt identification is thus made easy.

Other valuable features of the volume are:

Complete directions for the preparation and use of insecticides.   Spray formulæ, repellents, and fumigants described in detail.

Descriptions and photographs of spray machinery and accessories.

An illustrated discussion of the structure of insects.   How they live: their habits, senses, and manner of growth.

Insects as carriers of disease.   The typhoid fly, the malarial mosquitoes, and others.

The classification of insects, including illustrated descriptions of the various important groups into which insects are divided.

An account of the means by which insects are dispersed.

The natural enemies of insects.   How they are held in check by parasites, by fungous and bacterial diseases, and by birds and other larger animals.

How farm practice assists in insect control, showing the influence of plowing, cultivating, destruction of weeds, and the like.

A complete bibliography, giving an authoritative reference for each species treated in the book.

THE MACMILLAN COMPANY

Publishers   64-66 Fifth Avenue   New York

THE FARM WOODLOT

By E. G. CHEYNEY

Director of the College of Forestry of the University of Minnesota

AND J. G. WENTLING

Associate Professor of Forestry in the University of Minnesota

Illustrated.   Cloth, 12mo, $1.50 net

The whole subject of raising forests and producing timber as a part of a farming business is covered in this book.   Here will be found fully treated such topics as the rise of forestry knowledge in relation particularly to agriculture, forest influences, forest economics, the growth of the tree, the kinds of trees and the means of distinguishing them, the regeneration of the woodlot, the practical propagation of trees, methods of planting and thinning, the production of the forest, the best utilization of forests, the durability and preservation of timber.   There are also included tables of interest to lumbermen and a chapter on ornamental planting.   The volume is well illustrated, the illustrations alone largely explaining forest practices and making evident the differences in trees.

THE MACMILLAN COMPANY

Publishers   64-66 Fifth Avenue   New York

FARM STRUCTURES

By K. J. T. EKBLAW, M.S.

Associate in Agricultural Engineering, University of Illinois ; Associate
Member of American Society of Agricultural Engineers

Illus., Cloth, Crown 8vo, 347 pp., $1.75 net ; postpaid, $1.88

In the preparation of this book it has been purposed to provide a treatise concerning farm structures which will appeal not only to the teacher who desires to present the subject to his students in a straightforward and practical way, but to the progressive farmer who recognizes the advantages of good farm buildings. The popular literature on this subject consists mainly of compilations of plans accompanied by criticisms of more or less value, or of discussions of farmsteads too expensive or impractical to be applied to present ordinary conditions. The elimination of these faults has been among the objects of the author in the writing of this text.

The development of the subject is manifestly the most logical, beginning with a description of building materials, followed by a discussion of the basic methods employed in simple building construction, then presenting typical plans of various farm buildings in which the principles of construction and arrangement have been applied. Descriptions of the more essential requirements in the way of equipment and farm-life conveniences are appended. The illustrations have been prepared with the object of making them truly illustrative and of aid in the understanding of the subject matter which they accompany. Comparatively few building plans are included, since most building problems possess so many local requirements that a general solution is impossible; however, the plans presented are typical, and are so suggestive in presenting fundamental principles that a study of them will aid in the solution of any particular individual problem.

It is not intended that the study of this text will produce an architect; but it is hoped that it will provide the student with a sufficient knowledge of building operations to enable him, with some knowledge of carpentry, to erect his own minor structures and to differentiate between good and bad construction in larger ones.

THE MACMILLAN COMPANY

Publishers **64-66 Fifth Avenue** **New York**

" Every library, every country home, every city home with even the small-est window box, every school, and every office whose business in any way touches outdoor life, will find pleasure, profit, and inspiration in this great set of books."

THE STANDARD CYCLOPEDIA OF HORTICULTURE

Edited by L. H. BAILEY

With the assistance of over 500 collaborators. New edition, entirely rewritten and enlarged, with many new features; with 24 plates in color, 96 full-page half-tones, and over 4000 text illustrations. To be complete in six volumes. Sold only in sets by subscription.

Volume I (A-B), Cloth, 8vo, $6.00; Leather, $10.00

This work, an enlarged version of the famous Cyclopedia of American Horticulture, has been freshly written in the light of the most recent re-search and the most modern experience. It is not merely an ordinary revision or corrected edition of the old Cyclopedia, but it is a new work from start to finish, with enlarged boundaries both geographically and practically. It supersedes and displaces all previous editions or reprints of every kind whatsoever.

It is the fullest, the newest, the most authoritative of all works of its kind, and constitutes the most conscientious attempt that has ever been made to compress the whole story of our horticultural thought, learning, and achievement into one set of books. The text is under alphabetical arrangement supplemented by a Synopsis of the Plant Kingdom, a Key to the identification of the species, and an Index to the complete set.

THE MACMILLAN COMPANY
Publishers   64-66 Fifth Avenue   New York